GIS MODELING IN RASTER

GIS MODELING IN RASTER

Michael N. DeMers

New Mexico State University

JOHN WILEY & SONS, INC.

ACQUISITIONS EDITOR   Rayan Flahive
MARKETING MANAGER   Clay Stone
SENIOR PRODUCTION EDITOR   Patricia McFadden
SENIOR DESIGNER   Harry Nolan
PRODUCTION MANAGEMENT SERVICES   Hermitage Publishing Services

This book was set in Cheltenham by Hermitage Publishing Services and printed and bound by Courier Westford. The cover was printed by Lehigh Press, Inc.

This book is printed on acid-free paper. ∞

Library of Congress Cataloging in Publication Data:
DeMers, Michael N.
    GIS modeling in Raster / Michael N. DeMers.

ISBN 0-471-31965-1

Printed in the United States of America

10 9 8 7 6 5 4 3 2 1

PREFACE

GIS modeling has matured immensely in the past two decades, moving from once purely descriptive single objective types to more complex, prescriptive multi-objective ones. This is in no small way a reflection of the growing sophistication of the GIS user community. Today GIS modelers demonstrate an increased recognition of the capability of the software to model a wide range of domain types and complex spatial modeling situations. Combined with the rapid increases in computational speed, the availability of powerful, yet inexpensive GIS software within familiar operating systems on desktop computers, and massive increases in storage space; the use of GIS as a modeling tool is likely to grow at an even faster pace than in the past. As this pace quickens, so does the need to understand the underlying geographic concepts behind GIS modeling. In turn, as we learn more about how the current technology implements—or fails to implement—these geographic concepts, there is a concomitant increase in our need to assess the viability of our model output for decision making. For those more intrepid souls not content to accept the limitations of simplistic models, there is a desire to push the technology beyond its current state by developing new algorithms that more closely replicate the real world.

Unfortunately, many GIS modelers are being forced to learn their craft with little guidance. This is particularly problematic for those whose disciplines have only recently discovered the GIS toolkit, and for others who are already employed and lack a structured environment in which to couch their modeling tasks. Additionally, as more GIS programs are beginning to appear in university curricula, the need for this structured GIS modeling environment in the academic setting is increasing. This text fills these two complementary voids by providing an easy to understand structured textbook for college classes in GIS modeling and a useful reference for the practicing professional. For the college classroom it provides ample discussion questions and exercises that the instructor can use to enhance the learning process. For the working professional, it delivers examples of modeling methodology, together with detailed descriptions of such important topics as pattern recognition as a precursor to conceptualizing potential spatial models; Map Algebra as one common user interface to raster GIS; model formulation, implementation and verification; and advanced raster data structures for dynamic modeling.

For any who have attempted anything beyond the simplest modeling tasks with a GIS, it is obvious that no single text, no matter how well written or how brilliantly packaged, is not going to, by itself, make you a better modeler. The purpose of this book is to provide you with a sensitivity to the spatial components that comprise the building blocks of GIS models, a structure within which to model, and some insights into the common tasks that all modelers encounter, no matter what their domain of expertise. In short, it is meant as a modeler's companion. By employing current terminology it provides a language that modelers can share. Illustrating the commonality of modeling tasks among domains, it should assist in opening the lines of communication among disciplines. My intent is that an increased cross-disciplinary dialogue will result in still more and better GIS models and an enhanced understanding of the importance of geography and geographic concepts in an ever widening array of specialties.

To that end I entreat you, the reader, to discuss what you read in this book with your fellow modelers and/or students. Use the discussion questions at the end of the chapters or create your own questions. The examples I have provided are limited by my own experiences and should not be considered an end, but rather only the beginning of the process of learning how to model. By sharing your own modeling tasks with each other, whether in class, on the job, or at GIS user conferences, you will all become more insightful, more adept, and more proficient at GIS modeling. I hope you will share some of your insights with me as well so that my own students can benefit from your challenges, successes, missteps, methods, and models.

ACKNOWLEDGMENTS

Beyond my own efforts this book is the product of the industry of a large group of editorial and graphics professionals at John Wiley and Sons, Inc., as well as at Hermitage Publishing Services. My thanks to Ms. Carol Campbell for her assistance in producing the first drafts of many of the graphics. I am grateful to the outside reviewers who supplied many helpful suggestions. While production schedules may have limited the number of suggestions I could incorporate into the book, please be aware that I have considered them all. As always, I choose to assign them the credit for improving the quality of my manuscript and accept all responsibility for any omissions or errors in the text. Thanks are also due to Environmental Systems Research Institute (ESRI) and Earth Resources Data Analysis System (ERDAS) for providing access to software during the production of this manuscript. Finally, to my GIS modeling students who have had to endure the various stages of this work as it took form over these past several years, I offer my appreciation for your tolerance and your feedback.

CONTENTS

Chapter 8   Conflict Resolution and Prescriptive Modeling   162

Chapter 9   Model Verification, Validation, and Acceptability   175

## CHAPTER 1

Introduction

On completing this chapter and combining its contents with outside readings, research, and hands-on experiences, the student should be able to do the following:

1. List five tasks that extend the definition of geographic information systems (GISs) beyond the simple solution of geographic problems

2. Enumerate the primary reasons for the limitations of most organizational GIS implementations

3. List at least five discipline areas that could benefit from the application of raster GIS modeling

4. Describe at least five specific modeling tasks that could improve the quality and utility of existing GIS implementations

5. Define the advantages and disadvantages of raster data over vector types for GIS modeling

THE ROLE OF GIS MODELING

Geographic information systems (GISs) are computer software programs specifically designed to assist in the solution of geographic problems. But they are much more than that. They automate known geographic concepts and ideas; provide tools and justifications for geographic decision making; render explanations of distributional patterns of people, plants, animals, places, and things; and even predict new distributions and spatial arrangements through time. Even beyond these substantial tasks, within the hands of the innovative geographic information scientist the GIS can become an excellent laboratory for the exploration of the theory from which the GIS was originally derived. GISs allow scientists and social scientists to challenge existing observational hypotheses through detailed measurement, analysis, and geographic visualization (GV) of patterns that, without the ability of the GIS to deconstruct or aggregate, may yield false or misleading conclusions.

In short, the GIS allows both practitioners and theoreticians the opportunity to grab large chunks of the earth's surface and roll them around in their hands. It permits them to tinker with the landscape components of the earth, look at them sepa-

1

rately or in combination, strip away what is superfluous or what they don't want to look at, or add new or modified components to see its effects on other variables—all without the fear of disastrous and often irreversible consequences. As suggested by DeMers (2000a), the GIS gives us the opportunity to explore our world in much the same way that the geographers, naturalists, and explorers of the past did, but with a much more precise set of tools. More importantly, today's modern GISs, with their ready availability, advanced computational power, and vastly improved user interfaces, allow an immense array of practitioners to join in this exploration.

Although GISs have been available in one form or another since the 1960s, the 1990s probably accounted for the most rapid increase in their use to date. Beyond the vast improvements in the technology, one major reason for this increased popularity is the heightened recognition by a growing number of domain experts of the potential of the software to examine and to model the geographic components of their own problems. Domain experts include a wide array of environmental scientists, policy makers, crime analysts, urban or regional planners, health care professionals, engineers, military strategists, surveyors, oceanographers, agriculture professionals, landscape architects, academics, and many more. In fact the potential applications for personnel either using or likely to use GIS technology and employing geographic concepts are far too numerous to list. They do, however, all have one thing in common. They all share the need to examine mapped data relevant to their domains. They all seek answers to geographic problems. These problems may be simple, repetitive, and computationally exhaustive or they may be more complex, one of a kind, and computationally elaborate. In some cases, the problems may have no immediate exact solutions because of their complexity, vast data volumes, and context-sensitive nature of factor interactions. Or the problems may have no GIS solution at all because the theoretical foundations of the problems may not exist. This is when the GIS becomes an automated laboratory for hypothesis formulation and testing.

Whether oriented toward the application of existing GIS concepts and algorithms or directed to the development of new concepts, theories, and even the software itself, the primary purpose of GISs is to analyze geographically referenced data. The analysis most often takes place in the form of formal models of the human or physical environment in which the domain expert operates. These formal models can be designed to aggregate or disaggregate mapped data, to predict new distributions of mapped data, to define best locations for selected activities, or to describe the results of patterns of one variable on itself or on other mapped variables. If you are an application specialist, you must learn how to employ these models to obtain solutions to your problems and to provide justifications to decision makers. If you are a theoretician, you seek repeatable, quantifiable explanations of patterns by simplification within the modeling context, thus providing the framework for more exact, more accurate applications models. Finally, if you are a software developer, you must be able to create software that is both flexible and powerful to allow both the application specialists and the theoreticians to more effectively perform their tasks (Figure 1.1). In all cases, you must know how to model with a GIS (Figure 1.2). This is much more than knowing all the commands of a particular software package. Knowing which commands to issue the GIS software is helpful, but if you are unable to formulate a model prior to using the software, you will likely produce little that is useful. It is akin to being fully proficient with your word processor. Although this skill is important, it is not enough for you to be able to write a best-selling novel. Many organizations are initially frustrated with a GIS because it seems to add little to their goals and objectives. In a great majority of these organizations, especially when they are first exposed to GIS, the software operates primarily to store spatial data and information and to output the same data as hard-copy maps. This is most often not because the software does not contain the necessary algorithms to allow it to per-

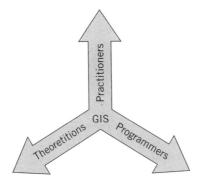

Figure 1.1 **A geographic information system (GIS) serves theoreticians, programmers, and practitioners alike.** An understanding of GIS modeling is important for practitioners, who will create the models; theoreticians, who develop the concepts of new models; and programmers, who must write to code to make the models work inside a GIS.

form the tasks asked of it, but rather that the software capabilities are not realized by its practitioners.

REALIZING GIS CAPABILITIES

With all tools, an inability to understand the tasks that they are designed for and that they are capable of performing limits both how well and how often they will be used. Just as carpenters need to know how different tools are most appropriately applied to different materials at specific times to produce specific products, GIS professionals need to know how the tools available in their GIS tool kit are best applied. If we continue the analogy, carpenters would not use a hacksaw to cut a two-by-four, nor would they use wood drill bits to drill a hole in sheet metal. Although the hacksaw could probably be applied to cut the two-by-four, the resulting cut would most likely not be as straight and even as desired, nor would a wood drill bit cut as easily through sheet metal as a metal drill bit that was specifically designed for the task. Most of us have, at one time or another, become frustrated when trying to perform simple plumbing

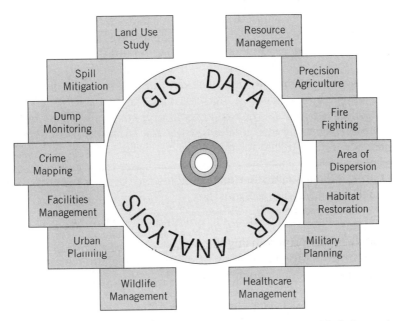

Figure 1.2 **The number of potential users for geographic information system data are many.** In some cases, a single spatial data set may be required for a wide range of tasks. This requires that care be taken in the development of the data sets.

tasks with a set of simple wrenches as opposed to the appropriate pipe wrenches and plumbing tools. Even if we eventually achieve the desired results, our willingness to perform such tasks again often wanes. This same reticence to pursue even simple GIS modeling tasks results if we are unfamiliar with the tools available to us.

A realization of GIS capabilities is more than a mere recognition of the computational power of the computer or the algorithms available in the software. In fact, it is quite often more important to be familiar with the task that you wish to perform. Like any tool kit, the sheer number of tools available can often be overwhelming. This is certainly the case with GIS where new algorithms and new approaches are being introduced almost daily. If, however, you know exactly what you are trying to model, what types of data are to be employed, and what the desired outcome or resulting model should look like, you are then able to select the appropriate techniques and algorithms. Through long experience and a trial-and-error approach, you might be able to struggle along until you serendipitously discover what works and what doesn't work, but most of us are not willing to wade through the nearly endless possibilities until we stumble along the right approach. It is both inefficient and likely to result in either poor or incorrect models. As with plumbers or carpenters who often go through a period as journeymen, learning from the experiences of others who are familiar with both the techniques of GIS and the modeling process itself will shorten the learning curve, enhance our capabilities, and provide us with the self-assurance we need to be effective modelers.

UNDERSTANDING THE MODELING PROCESS

As suggested by the above heading, GIS modeling is a process. This process requires a different way of thinking about the world than many of us are accustomed to. First, we need to be aware of how we represent the world in a GIS. This is particularly important in a raster environment where we divide our study area into grid squares (usually). More importantly, however, it requires us to examine our data and to think about the spatial components contained within them. This spatial thinking is critical to GIS modeling. I cannot emphasize that enough. Before we can use a set of tools that is uniquely designed for spatial modeling, we must first be able to define how we are to abstract geographic space for later input into the GIS. We must also be aware of the possible spatial relationships that we are searching for, what these spatial relationships may or may not tell us about our environment, and how we can best enumerate, measure, categorize, and combine them to achieve meaningful models.

This process can be confusing without a structure within which to operate. This structure begins by first deciding what our model is to do. We ask ourselves such questions as:

- Am I creating a database that will allow me to ask questions about what things are where?

- Is my model going to help me quantify an existing pattern so I can better understand it?

- Is my model attempting to examine multiple relationships from different maps?

- Am I making a model to show how things change through time?

- Is my model designed to predict something?

- Am I attempting to find the best places, routes, or scenarios for some form of activity?

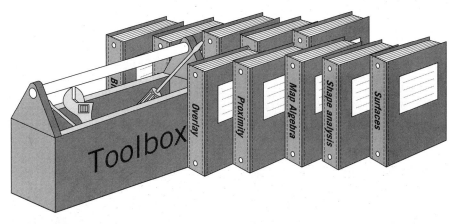

Figure 1.3 Geographic information systems can be thought of as a toolbox with many different techniques available for a wide range of simple and complex tasks. As with any set of tools, it is important to know the best tool for the job before beginning.

By defining what it is we are trying to say about our data, whether we are trying simply to explain existing relationships, to predict consequences from such relationships, or to define the best situations for appropriate uses of our environment, we have the most important component of our model. Once we know what we want the model to tell us, we have the basis for defining what types of data we need to acquire to build our model (our basic building materials), and we can begin creating a design for how the model will be put together (a blueprint), and this in turn will help us decide how best to apply our GIS algorithms (our toolbox) to achieve the desired outcome (Figure 1.3).

This approach has proven successful for a wide array of complex models. It allows us to approach the model as a set of individual components or modules, each of which will be individually examined as a simple model. Once we have all the necessary data, a detailed blueprint, and appropriate algorithms for each component, we will then link them into a larger, more complex model of our environment. In this way, we use a divide-and-conquer approach to model building. By so doing, we are able to simplify the problem so that it does not become overwhelming. Additionally, we can add components later on if we discover that something is missing. In fact, this modular approach is ideally suited to building models that continue to grow as our knowledge of our environment grows. By building up simple, easy-to-understand modules, we gain confidence in our ability to use the GIS toolbox effectively. This confidence is nearly always translated into a willingness to use the GIS for more than map storage and retrieval.

WHY RASTER GIS MODELING?

At this point, it is important to state clearly that GIS modeling is not restricted to raster data types. In fact, many very sophisticated and elegant models are done in vector. There are also some things that vector GIS can do much better than raster, especially where networks and polygonal data are concerned (DeMers 2000a). So why, you might ask, are we only examining GIS modeling in raster? One reason is that it is difficult to discuss GIS modeling in both raster and vector simultaneously. The algorithms that are used frequently differ substantially between the two general data structures, and this detracts from the clarity of the discussion. With the availability of relatively inexpensive commercial desktop vector GIS packages, and with the result that the vast majority of introductory-GIS students become exposed to these packages first, raster modeling is now often relegated to the end of these courses. This is in contrast to previous times when inexpensive raster packages, although not often commercial packages, were the norm and vector packages were either unapproachably complex or prohibitively expensive. These are purely pragmatic reasons for why this book covers solely raster modeling.

There are, however, other reasons that this book explicitly covers raster modeling. With the increased availability of both inexpensive and commercial-quality raster GIS packages, there has been an increase in the modeling capabilities they provide as well. Compare, if you will, Tomlin's (1983) original Map Analysis Package (MAP) with the new generation of products, such as ArcGrid, ArcView Spatial Analyst, GRASS, Blackland, ERDAS Imagine Spatial Modeler, IDRISI, PC Raster (van Deursen 1995), and a host of others. These packages offer far more algorithms than did MAP, including, for example, integrated remote sensing capabilities (e.g., GRASS and ERDAS Imagine Spatial Modeler), advances in neighborhood and map calculator functions (e.g., ArcGrid, ArcView Spatial Analyst, SPANS), explicit dynamic modeling (e.g., PC Raster, Wesseling et al. 1996), true three-dimensional modeling, and even fuzzy logic (e.g., IDRISI).

Beyond the simple increase in power that these new raster products provide over their previous relatives, raster data structures also provide an increase in flexibility of modeling of surfaces beyond the vector models such as the triangulated irregular network (TIN). Where environmental modeling includes such functions as hydrological flow modeling, travelshed modeling, and even overlay modeling, the raster GIS has far more options than most vector equivalents. Where models are designed to describe diffusion or dispersal events such as those involving aerosols, dust, fire, and/or disease vectors, the raster GIS excels in its ability to address these types of surface-dominated questions.

Another reason for us to address raster GIS modeling is the explosion in availability of raster data sets, particularly those available from airborne and space-borne sensors. These types of data are particularly well suited for tasks related to monitoring large areas and for updating existing information. The increased resolution and decreasing costs of these remotely sensed data make it even more attractive to combine image-processing algorithms with raster GIS algorithms to produce a suite of techniques that extend the capabilities of both software package types. This also explains why remote-sensing and GIS software vendors have been cooperating to make their data structures compatible (cf. ArcView Image Analyst).

Another good reason for focusing on raster modeling is that the heart of the GIS modeling framework, called cartographic modeling (Berry 1993, Tomlin 1991), that has become the standard methodology was first implemented inside a raster GIS—the MAP (Tomlin 1983). Because MAP is probably the single most copied GIS model available, this approach is among the most common. Recent advances in raster GIS software have frequently adopted this standard approach as well as its terminology. Both because the published literature is replete with this terminology and because the cartographic modeling methodology has become a "best standards and practice" within the GIS modeling community, it is both practical and advantageous to employ it here.

WHAT THIS BOOK IS ABOUT

This book then is about cartographic modeling. It will assist you in understanding the terminology of cartographic modeling and the implications of the terminology. In addition, it will take you through the complete process of cartographic modeling. But it is more than a textbook on cartographic modeling. Although cartographic modeling is at the core of GIS modeling, it does not preclude the addition of other techniques that are not commonly considered part of its domain. We will examine a superset of both GIS and non-GIS modeling techniques, look at different types of raster data GIS models as they are applied to modeling, and examine topics that are often considered to be only loosely connected to cartographic modeling.

In particular, we will examine in detail the methods of raster GIS modeling beginning with defining our problem in spatial terms, selecting the appropriate raster data model(s) for the model, and defining complementary technologies that can assist us during the modeling process. We will examine a variety of GIS model types to allow us the opportunity to become comfortable selecting the best approach for each. For each model type, we will spend a large amount of time on identifying model components and flowcharting the model so that we begin to feel comfortable with the modular approach.

Beyond just building models, we will also examine the model implementation. This means we will actually see what happens when we run the model we have created. This will allow us to verify that the model is performing the way it should, both computationally and conceptually. Model verification will also require us to examine whether the models we create are acceptable to the potential user. Beyond model verification, we will also discuss the methods available to us to validate our model, from reverse computation to expert opinion and the use of validation data sets.

Additional modeling topics will involve the application of conflict resolution, particularly where spatial conflicts within a GIS model are involved. These techniques will give us some insights on how GIS is actually used and how conflicts arising from disparate demands on limited land resources can be addressed within the GIS. This book will also examine the idea of time in GIS at both the conceptual and the practical level. Although most raster GIS data models do not explicitly incorporate time, it can be managed. And we will look at some of the nontraditional raster approaches to space–time modeling.

Beyond these modeling topics, we will also give some time to alternative logics, especially the application of fuzzy logic to raster GIS modeling. Such academic topics necessitate that we go beyond the "button bar" on the GIS, and this text will present a brief examination of computational geography from both the perspective of computer programming using macro languages and from the perspective of the spatial analyst who sees problems that are not currently being addressed either by the GIS modeler or the GIS programmer.

Recognizing that this text is likely to have a wide audience, you should feel free to select which topics interest you, especially the more advanced topics. GIS practitioners, for example, may not have an interest in programming but can concentrate more on the chapters on modeling. Those familiar with the various raster data structures might feel comfortable just perusing the second chapter. In other words, like GIS modeling itself, you are free to take a modular approach to the text as well.

Chapter Review

Geographic information systems (GISs) automate geographic concepts, assist in decision making, help explain distributions, and can assist in hypothesis formulation and testing. These tasks can be applied to a wide range of both practitioners and theoreticians by allowing them to manipulate portions of the earth that are stored as map data in the computer. The current popularity of GISs is in the multitude of domains in which they can be applied and in their ability to automate simple but repetitive map-based tasks as well as complex ones. A primary reason that organizations often become frustrated with a GIS is that they lack a thorough understanding of both the GIS modeling capabilities and the applicability of their data to spatial models.

An important first step is to view GIS models in a modular form, where simple models are combined to create larger, more complex ones. One begins by first deciding what the questions are that we wish to ask of the GIS. This helps us understand what the final product is that we want to create. By working backward, we can separate the model into its necessary components. Each component is then separated into the exact data that we need to build the model.

Although GIS modeling is available for both vector and raster data structures, we will exclusively examine raster modeling. This is done partly to avoid the complexity of examining both raster and vector models simultaneously. It is also done because of the ready availability of raster data sets, the easy application of existing raster modeling terminology and conventions, the ability to conveniently model surfaces in raster, and because there are more and more professional raster GIS packages available.

This book is about cartographic modeling, but it includes much more. It incorporates different raster data types, links to non-GIS modeling, and shows how GIS modeling can eventually lead the student beyond the existing software limitations into macro modeling. A primary focus of the text is on spatial analysis and modeling rather than on how a particular software package is used to model geographic phenomena.

Discussion Topics

1. Suppose you have just been hired as the senior geographic information system (GIS) applications specialist for a regional environmental management agency that is new to GISs. Although you have several competent software technicians under your directorship, their training is mostly with non-GIS software and they are generally unaware of the computational power of the GIS. Additionally, they have little feel for the spatial data needs of the agency. In the 3 years prior to your starting work, the primary tasks the GIS has been applied to are inputting existing maps and providing hard-copy output of those same maps to the company field agents, who are mostly biologists with no GIS training. Suggest how you would approach your superiors to suggest ways the GIS could be used. In particular, suggest the types of GIS operations that would be useful to the field agents, what training programs might be necessary for them and for the technicians, and how both of these might be useful for enhancing the utility of GIS in this work environment.

2. You are involved in a local speakers club, and two or three of your club members notice a copy of this book in your briefcase. They ask you to define what a GIS is and to describe what it is used for (i.e., what it does). Rather than answering them individually, prepare a 7-minute lecture for the club at large that does this. What would you include?

3. A friend of yours has been working with vector GIS for several years and is quite content with its modeling capabilities. She/he asks you why you would bother working with raster GIS modeling when the vector provides a much more realistic set of output maps. Describe for this person the advantages and disadvantages of raster over vector GIS modeling.

4. You have been asked to establish a regionwide GIS coordinating body. Among the members of the committee that you are organizing is a wide variety of non-GIS professionals. Although they have heard of GIS and know that it has something to do with computer mapping, they don't quite understand how it could be used for their particular endeavors. Provide the committee with a list of at least five separate subject domains that could benefit from GIS analysis and provide one or two

concrete examples (perhaps from GIS Web pages with which you are familiar) of how it has already proven beneficial for each. Besides just showing how it has been used, provide specific measurable benefits.

Learning Activities

1. This chapter extends the definition of a geographic information system (GIS) to include more than just a piece of software for solving geographic problems. Create a scrapbook that lists each of the general modeling capabilities that will be examined in this text on a separate page (or pages). For each capability, clip articles and GIS output from newspapers, trade journals (e.g., *GIS World*, *GeoInfo Systems*, *GIS Europe*, *Business Geographics*, *ArcNews*), newsletters, Web pages, and any other sources you can find that illustrate these capabilities. Next to each article, provide a brief description of what you are illustrating. Note: You may want to use a looseleaf scrapbook because this information becomes very useful as sources of new ideas as your modeling skills grow.

2. Select a set of subject domains that you are interested in. For example, if your primary interest is in environmental modeling with GIS, subdivide that into such things as monitoring of U.S. Environmental Protection Agency Superfund sites, wildlife habitat analysis, hydrological modeling, environmental planning, and preparation of environmental impact statements. Once you have selected your set, begin compiling a bibliography of research articles and books that illustrate the modeling domains with which you are interested.

3. While you are compiling your research articles and books, create a table (either manually or with a spreadsheet) that lists a short reference (i.e., *Name and data*) down the lefthand side. On the top, list at least the following basic GIS analysis operations: (1) simple enumeration, (2) simple measurement, (3) single-map comparisons, (4) overlay operations, (5) surface operations, (6) predictions, (7) prescriptive models. You can make as long a list as you want. Now, for each article you find, check all the analysis operations that apply. This will prove very enlightening as you discover what others are doing. You can also apply this to the scrapbook items you collected for activity 1 as well.

## CHAPTER 2

Nature of the Data

On completing this chapter and combining its contents with outside readings, research, and hands-on experiences, the student should be able to do the following:

1. Define and explain the term *quantizing* as it refers to geographic space

2. Explain the ramifications and implications of spatial quantizing as it relates to both the representation and the modeling of geographic phenomena

3. Diagram and explain how points, lines, and polygons are represented in raster format

4. Explain what surfaces and fields are and provide concrete examples

5. Diagram how surfaces and fields are represented in raster tessellations

6. Describe the four basic raster data models (simple raster [all variants], extended raster, quadtrees, and cellular automata)

7. Explain the difference between cellular automata and other types of raster data models

8. Enumerate the advantages and disadvantages of raster data representations over vector representations both in terms of data storage and modeling

9. Describe what types of models raster GISs are more usefully employed to solve and provide concrete examples of this

10. List several sources of raster data for GISs, including, but not limited to, remotely sensed data sources

11. Explain the importance of grid cell size, projection, and grid system as they relate to raster GIS modeling

12. Provide some basic sources of raster data error, including generalization, classification, interaction, and entity–attribute error components

13. Describe, in general terms, how the temporal component is handled in traditional raster GIS operations

INTRODUCTION TO RASTER TESSELLATIONS

It was suggested in the first chapter that raster data are both commonplace and very effective data types for GIS modeling. Although some of you may be very comfortable with this type of data and its capabilities, it is important that we review these general concepts and examine them in somewhat more detail. This provides us with common ground and a common terminology on which our modeling vocabulary and concepts can be built.

All raster data types are tessellations, or ways of dividing our geographic space so that it can somehow be represented inside the computer. The process begins by conceptualizing our real world and then converting it to a cartographic abstraction of that reality (Figure 2.1). Once completed, the cartographic product is then converted to a digital equivalent through some form of tessellation. The raster tessellation divides geographic space into a series of discrete chunks within which to represent real geographic data. This approach is called quantizing space (Kemp 1993). By that, I mean that we divide the spatial data into quanta or packets on which we will perform our analytical operations either individually or collectively (Figure 2.2). This type of representation converts both continuous and discrete spatial data into discrete units on which the software operates. It provides somewhat less exact locational information than its vector counterpart, but it adds the ability to store the various types of spatial entity information uniformly. As the amount of area on the ground represented by a single grid cell increases, the locational accuracy decreases. Stated differently, as the grid cell resolution decreases, the locational accuracy decreases (Figure 2.3). Although not restricted to a particular shape, grid cells are nearly always represented as squares. Other shapes are possible—for example, parallelograms and hexagons (Figure 2.4). Although these shapes have their own advantages over the square, such as their applicability to representing surface data (e.g., triangles) or spherical objects (hexagons), or remotely sensed data (parallelograms), the simplicity of the square, the ease of operation with the square, and the often intuitive nature of the square make it a preferred raster tessellation. Most of the original

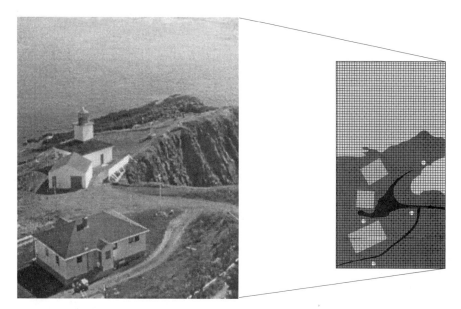

Figure 2.1   Whether analog or digital, the map is an abstraction of reality, using symbols to represent the objects on which we will base our geographic information system models.

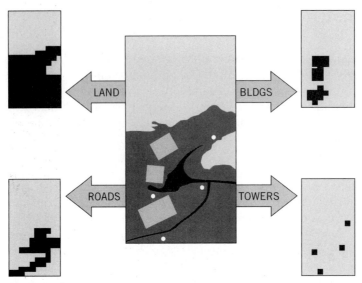

Figure 2.2   The raster representation of point, line, and area features involves dividing our geographic space into discrete quanta that we call grid cells.

raster-based GIS software was developed using the square grid cell because of these properties, and the most commonly copied raster GIS—Tomlin's (1983) MAP—took full advantage of this tessellation. This latter may also help to explain, to some degree, the predominance of this grid cell type for GISs.

The raster or grid cell is designed to represent known or perceived geographic objects and provides mechanisms to store the descriptive information about those objects as well. Traditionally, geographic space has been defined by several types of geographic entities (objects). The simplest are points, lines, and areas (polygons). Points are represented by a single grid cell whose coordinates are most often a function of their relative location in a matrix of grid cell locations. In other words, its location is relative to all other grid cell locations and is most often identified by a set of X and Y locational coordinates in Cartesian space (Figure 2.5a), hence the relative lack of locational accuracy. Most modern GISs provide for a direct linkage between the Cartesian coordinate space and a geographic coordinate system, thus allowing

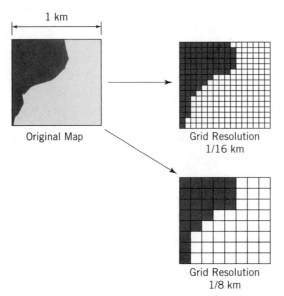

Figure 2.3   As the grid cells get bigger (as resolution decreases), the locational accuracy decreases as well.

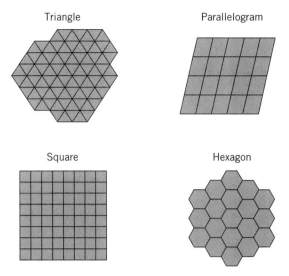

Triangle Parallelogram

Square Hexagon

Figure 2.4 Although the square grid cell has become fairly standard, there are other forms of tessellation of geographic space that could be applied. Each has its own unique properties and utility.

for geocoding, for overlaying raster coverages stored in different projections, for performing projection changes, and for doing edge matching, and permitting various other spatial manipulations such as rubber sheeting. These topics are beyond the scope of the current volume, and the reader is directed to other sources for more information (cf. Chrisman 1997, Heywood et al. 1998).

Extending the raster representation of geographic objects to lines and areas is simply a matter of adding collections of raster point locations, together with the line occupying the intervening space. For example, a line in raster is a linear collection of grid cells, the location of each cell defined, as before, as its relative location in the overall matrix of grid cells (Figure 2.5b). In theory, the line represented by the collection of grid cells is one-dimensional, having length as its only measurable spatial

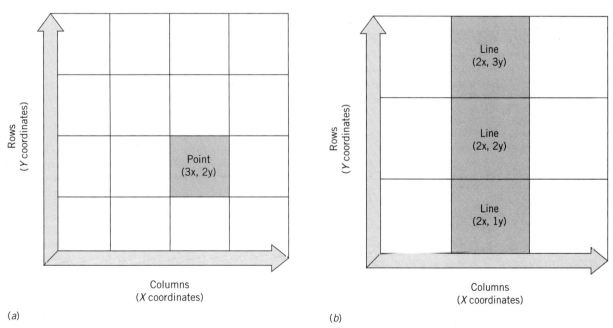

(a)

(b)

Figure 2.5 In Cartesian space, grid cell locations are located by their position as ordered column and row values. Thus, points (a) are represented by a single coordinate pair, lines (b) by linear sets of coordinate pairs, and areas (c) by groups of coordinate pairs. *Figure continues.*

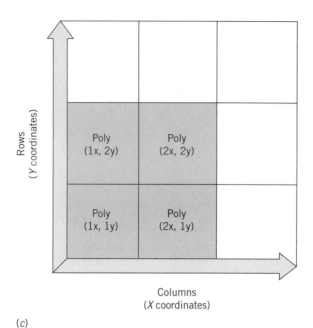

(c)

Figure 2.5 *Continued*

dimension. Of course, grid cells, by definition, occupy two dimensions, but we suspend our disbelief to allow this abstraction. By extension, then, the raster representation of areas or polygons is a two-dimensional collection of grid cells (Figure 2.5*c*), each cell's location, as before, determined largely by its relative position in the matrix, together with any grid system that is superimposed on the matrix.

Although the raster tessellation is less accurate in absolute geographic space, its consistent, regular shape makes it easy to compare the contents with other grid cell–based thematic maps. In addition, it also allows all of the basic geographic objects (points, lines, and polygons) to be represented by the same tessellation. Perhaps even more important, it permits the representation of the final type of geographic object—surfaces or fields, again with the same tessellation model.

Surfaces, the final major geographic objects needing to be represented inside the GIS, are based primarily on the idea of the statistical surface. By this I mean that surfaces need not be elevational or topographic in nature. Rather they can represent any set of data that is, or can be assumed to be, continuous and that is measurable on an ordinal, interval, or ratio scale of data measurement (Robinson et al. 1995). The concept of the field is an extension of the statistical surface but includes any statistical surface-related data that can be represented as an equation. In fact, it is the representation of the statistical surface as an equation that is the most important aspect of the field. However, fields can also include such statistical surface representations as gravitational attraction between/among economic institutions, resource sources and sinks, spatial transition probabilities, and predator–prey interactions (Hilborn 1979). Many of these statistical surfaces can be represented by a single, although complex, equation, but this does not preclude our ability to represent them as discrete packets of information in raster. Given our focus on raster tessellations, we will consider surfaces and fields only within that context.

As with point, line, and polygonal entities, surfaces can also be represented by the quantizing of geographic space. If, as we have already stated, we assume that statistical surfaces are composed of continuous data, then their abstraction to raster involves converting the continuous nature of the surface to a set of discrete tesserae … what we have already called grid cells. Given that surfaces contain three dimen-

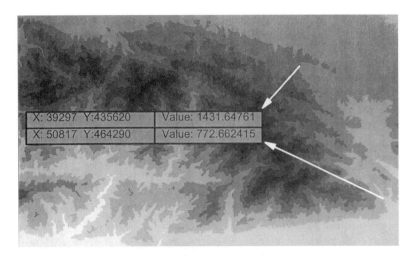

X: 39297 Y:435620   Value: 1431.64761
X: 50817 Y:464290   Value: 772.662415

Figure 2.6 The representation of surface values in raster involves assigning a single elevation value to each discrete grid cell. In turn, the grid cell's location is still recorded as a single pair of column and row values.

sions—length, width, and height—the relative nature of the locational information contained within the representative grid cells is extended to the third dimension. As such, there is a loss of spatial accuracy in the X, Y, and Z dimensions. Again, however, this loss of locational accuracy is counterbalanced by the ease with which the surface information can be manipulated, analyzed, and compared with other coverages.

Representing statistical surface data in the raster tessellation model normally involves providing a single Z or elevational value for each X and Y grid cell location (Figure 2.6). The result is that the Z value will be representative of some location within the grid cell but will be illustrated by the entire area occupied by the grid cell. As such, the larger the grid cell, the less accurate the Z value will be for the area. This single-value-per–grid cell approach is a common approach to representing statistical surfaces in raster and represents a typical raster data model used to represent all geographic entities. The next section discusses some alternative approaches to this traditional approach and shows how extending this model can have some useful GIS modeling consequences.

RASTER DATA MODELS

The tessellation is primarily designed as a graphic representation of the entities, and the attribute assignments are typically linked to them through explicit numerical assignments of each grid cell. For example, a point object represented as a raster is usually assigned a single numerical value to represent it. Perhaps a value of 2 could be used as a nominal code for representing telephone poles as point objects. This is similar to the assignment of digital number values (the attributes) ranging, for example, between 0 and 255 for remotely sensed data represented as pixels (picture elements). Although this approach is easy to understand, it is only one of several ways that the data can be represented, and I will refer to it here as a simple raster data model. The idea of a data model is that we need to formally construct a method the computer can use so that the graphical entities and the descriptive attributes can be linked, particularly where multiple themes are used. We will begin by examining the simple raster and then move on to more complex and/or more exotic approaches to modeling raster GIS.

Simple Raster

In the simple raster data model, for each location in our raster grid matrix, there is a single numerical value representing any of the point, line, area, or surface features we encounter in the real world. Its purpose, like all the other approaches you will learn about, is to allow the modeling process to take place. This takes it far beyond the simple coding of entities and attributes. For modeling, it is essential that grid cells within a single theme be able to interact with others in the same theme and for them to interact with grid cells in one to several additional themes. Without this property, the systems would allow for the output of raster maps but would be severely limited in their ability to model. Among the ways to structure such models is a general class of raster data models that we will call simple because they store a single value for each grid cell for each theme; they are still prevalent in both education (e.g., the original MAP and its variants) and professional (e.g., GRASS and IDRISI) raster GIS packages.

Within this general class, there are, however, several variants to the model, each of which approaches the access to data and information contained in the raster tessellations differently. The first of these is called the MAGI model (DeMers 2000a), which addresses each grid cell in a single theme individually and, when necessary, makes comparisons to other grid cells contained in different themes in columnar fashion (Figure 2.7). This approach was among the first developed and shows an early, necessary focus of GIS to compare and contrast multiple thematic data for modeling. Although effective, this approach is not as intuitive as one might like, especially given the tendency of GIS modelers to view themes as a whole in at least two dimensions, rather than viewing each grid cell as part of a column.

The IMGRID variant, unlike its MAGI predecessor, uses the two-dimensional array or theme as the primary unit on which queries are to be performed (Figure 2.8). When this system was first devised, computer memory and storage were at a premium. To allow for this, each theme was very specific requiring that the variables be binary. This necessitated that all theme categories be coded as either 0 or 1 values. In this way, a theme like land use would be unworkable because it could contain more than two categories, thus disobeying the binary rule. Instead a theme like land use would have to be simplified. Some examples of legitimate theme categories would include the following binary pairs: land/water, urban/rural, industrial/nonindustrial, polluted/nonpolluted. As you can see, the binary approach simplifies the coding to 1 and 0, thereby saving space in the computer. However, if you consider our statistical surface representation, there is no way to include such nonbinary data as topographic values, friction surfaces, impedance values, gravitational attraction values of any kind, probability values, and the like. Additionally, you might easily imagine as many as 100 categories of land use or land cover for a single map. If this binary approach were taken, there would need to be dozens of themes created to

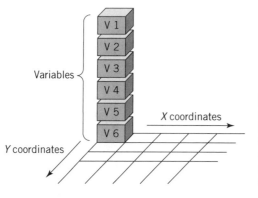

Figure 2.7 The MAGI method of representing raster geographic information system data addresses each value as part of a column. This provides a vertical linkage among multiple thematic grids but makes addressing horizontal groups of cells less efficient.

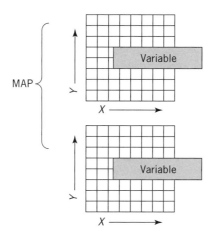

Figure 2.8   The IMGRID method of raster representation is a horizontal model requiring that each theme be represented in such a way that 0 and 1 can represent them. In this way, each theme must be Boolean, allowing no gradation and no multiple categories within the same set of horizontal grid cells.

cover just this single conceptual theme. Indeed, if each land use category were to be represented with a value of 1 and its alternative (a "not" category, if you will) were to be represented as a 0, then an analog map with 100 land use categories would require 100 themes to store all of this information. Still, before you dismiss this model, the idea of binary themes or binary maps will be revisited within the context of modeling as a means to account for large numbers of variables and themes.

The final variant of the simple raster data model is that developed by C. Dana Tomlin (1983) for the MAP as part of his doctoral dissertation at the Yale University School of Forestry. The MAP data model, still among the most-copied raster data model in the world, addresses each theme collectively, thus allowing multiple categories of thematic values within each theme or coverage. In this way, the theme is the primitive element on which modeling takes place, and statistical surfaces and fields can be easily included within the GIS, just as points, lines, and areas are (Figure 2.9). The original data model created by Tomlin used a "single value per cell" approach, as do all the other variants of the simple raster data model. This approach is much more compact than its predecessor (IMGRID) but still complicates the storage of very complex categorical information that might easily be contained in a single category. Take as an example the single category of row crops that might be included in an agricultural theme. Within row crops there may be additional information. The list might include some of the following possibilities (crop type, variety, date planted, pesticide amendment, fertilizer type, potential yield). This suggests the same data explosion problem as that encountered with the IMGRID data model. It further suggests that the simple raster model needs to be extended to diminish this effect.

Simple raster models and their associated tessellations most often encode integer numbers rather than rational (decimal) numbers. Although it is possible to store rational numbers using simple raster GIS models, the amount of storage necessary to

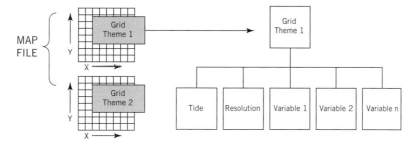

Figure 2.9   C. Dana Tomlin's MAP (Map Analysis Package) stores grid cells in such a way that each theme can be addressed individually. In addition, one or more of the thematic categories can also be addressed at will.

hold them and, more importantly, the amount of computational resources required to model with rational numbers often exceeds those available in even the most powerful workstations. This has suggested to some that GIS be ported to the parallel processing capabilities of supercomputers, but the relative lack of availability of supercomputers most often precludes this approach beyond purely experimental uses within research. As such, it is impractical for applications and commercial implementations of GIS.

Extended Raster

The simple raster model causes data explosion because all categories must be explicitly encoded for each grid cell for each of its themes or coverages. The extended raster model is actually an extension of the MAP model in which the theme is the primary feature that is addressed. This feature of the MAP model is what makes it work for the extended raster model because the extension addresses the theme as the basic entity unit and extends it by allowing multiple thematic data for each grid cell. In this way, the basic theme or coverage is addressed first (e.g., we locate and isolate the theme *forests*). Within *forests*, we have a set of grid cells grouped by forest categories such as *white pine, blue spruce, northern red oak, trembling aspen, ash,* and so on. For each of these categories, we have a set of grid cells whose categories are explicitly encoded as numerical values, just as they would be under the MAP model itself. Then, however, each category has linked to it a set of tabular data contained in a relational database management system (Table 2.1). For example, with our *forest* theme in Table 2.1, there are additional descriptive data, such as canopy density and percentage insect damage. You also note that this allows for the storage of additional descriptors such as the value associated with each number and a count of grid cells (the equivalent of the percentage of the map covered by each category). As you can see, this allows a great deal more attribute data to be included for each theme, thus saving computer space by not requiring additional themes for each category, and also putting this thematic data at the fingertips of the user. This latter is very important to the modeling process because as each map is used, its associated attributes are carried with it. Additionally, as each map interacts with each additional map, the extended thematic data are also carried along to the newly created themes.

2	1	4	4	4	1
2	2	0	5	5	1
2	2	1	5	5	1
1	2	4	1	2	1
3	3	3	1	2	1
1	1	3	0	0	4

TABLE 2.1 Extended MAP Data Model

Value	Count	Type	Canopy Density	% Bug Damage
0	3	No data	No data	No data
1	12	White pine	30	8
2	8	Red oak	65	10
3	4	Blue spruce	10	0
4	5	Aspen	45	20
5	4	Ash	80	35

The extended Map Analysis Package (MAP) data model involves the creation of attribute table entries for each grid cell represented. This allows a single grid cell to have multiple descriptors without requiring additional thematic grid sets.

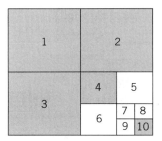 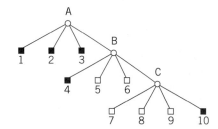

Figure 2.10 **Quadtree representation of grid cells divides the geographic space into successive quadrants.** This approach to raster data representation stores uniform (homogeneous) square groups of cells as a single value or quadtree level. In this way, highly homogeneous regions will normally be stored efficiently.

Quadtrees

In all the types of raster data models we have encountered so far, the basic assumption is that each grid cell or raster occupies the same amount of geographic space. Although this tessellation is relatively easy to understand and certainly among the most commonplace, it requires the storage of each grid cell and each associated set of attributes (in the case of extended raster) as a separate item or set of items. This results in a vast amount of unnecessary storage. For these traditional tessellations, there are some methods of encoding and storage that reduce the quantities of data stored. For example, run-length encoding and block encoding (DeMers 2000a) both save substantial amounts of storage space by aggregating whole regions of the earth as single units. In some cases, the GIS software associated with these techniques requires that the data be converted to their noncompact forms prior to analysis and modeling, and in others they do not. Essentially, however, these compact methods are designed for input and storage rather than for modeling. Another compact model, designed originally for artificial intelligence and expert systems work, has been adapted to operate as a form of raster GIS model. This model, called the quadtree, divides the earth into successively smaller, homogeneous square units (Figure 2.10) and allows the user to decide how detailed a tessellation (called a quadtree level in this case) he or she might wish to use for modeling. In addition, the quadtree data structure is readily usable for modeling without reconfiguration to its most primitive spatial level first. Two primary software packages are available that use this data model—one, a noncommercial version from the University of Maryland called Quilt (Shaffer et al. 1990), requiring the development of a user interface. The other, called SPANS, is a professional, commercial package that is well suited to the ready development of raster GIS models. Fortunately, the approach to modeling is much like what one might encounter in any of the more traditional raster GIS packages.

Cellular Automata

A more recent entry into the world of raster GIS models is actually a parallel development originally conceptualized by John von Neumann and his friend S. M. Ulan, and operationalized to model life itself as part of the "Game of Life" (Gardner 1970, 1971). Called cellular automata (CA), these data models are also based on the uniform grid cell tessellation of geographic space. And, like the simple raster GIS model, the cellular automata stores a value that represents some attribute of the land surface (Theobald and Gross 1994). Unlike typical raster GIS, however, the cellular automata incorporates an explicit set of transition rules, defined by the modeler and

Von Neumann Neighborhood   Moore Neighborhood

Figure 2.11 The two most common neighborhoods used in cellular automata representation of geographic space are the von Neumann neighborhood, which recognizes only adjacent cells, and the Moore neighborhood, which recognizes only diagonal cells.

designed to allow for dynamic modeling. These two characteristics are why cellular automata are often called tessellation automata or even iterative array when external inputs are allowed.

The actual definition changes from author to author, especially with regard to how the transition rules within the grid cells are defined. The definitions range from grid-based models with either discrete or continuous state values, with either stochastic or deterministic state transitions rules, or with either synchronous or asynchronous state changes (Childress et al. 1996). In some cases, these transitions rules may have been derived through linguistic acquisition of model heuristics, or rules of thumb (Wu 1996); in others, the rules are more closely linked to actual environmental conditions (Childress et al. 1996). Some have even attempted to link cellular automata with existing raster-based GIS through modifications of Tomlin's Map Algebra (Takeyama and Couclelis 1997). We will discuss more of the modeling aspects of these data models when we examine the use of GIS for modeling temporality later in this chapter.

To perform its modeling, the cellular automata model requires that the attributes be recorded as rational numbers. This is necessary so that the often subtle transition states can be represented by equally subtle changes in the fractional numbers. Additionally, the cellular automata model relies heavily on the concept of the neighborhood—a concept we will revisit in Chapter 4. The two most common neighborhoods are the von Neumann neighborhood (adjacent neighbor cells) and the Moore neighborhood (diagonal neighbor cells) (Figure 2.11) (Childress et al. 1996, Hogeweg 1988). Both of these are immediate neighborhoods: that is, all of their cells are immediately adjacent to or diagonal to the central cell (we will later call this the target cell). Although these are the most common neighborhood configurations, the cellular automaton is not limited to immediate neighborhoods. Whether immediate or extended neighborhood cells, one really important characteristic of cellular automata is that they allow an oportunity to study emergent global properties and behaviors by understanding the local processes only (Theobald and Gross 1994). Among their primary limitations are those shared by most grid cell–based environments—the assumptions of regularity, homogeneity, universality, and closure don't necessarily correspond effectively with real life.

ADVANTAGES AND DISADVANTAGES OF THE RASTER APPROACH

Regardless of the grid data models used, most of the disadvantages stem more from the properties of the tessellation itself rather than the specific model within which they operate. In particular, the primary disadvantages are related to the relative lack of spatial resolution compared with their vector counterparts. From a purely carto-graphic representational viewpoint, vector tessellations are far more like their ana-log cousins than are raster types. For this reason, some people feel more comfortable working in a vector environment, particularly those that primarily produce carto-graphic output and whose modeling is less complex or more graphical—that is, more typically one of comparing cartographic objects within a single theme or comparing cartographic products. Tomlin (1990) indicated that raster GISs are more position oriented than their vector counterparts, which are more theme oriented, because of their strong reliance on the polygon as a major data type.

This may seem counterintuitive, given that vector systems assign specific coordi-nates to individual points as well as to the points that collectively comprise lines, polygons, and TIN-based surfaces. Although this is true, the intermediate space between sets of points is effectively an implied space rather than an explicit space. There are no explicit coordinate representations for the intermediate space. In raster systems, because all the geographic space is occupied to one degree or another by grid cells, the only unaccounted-for space is that associated with the quantization of the geographic space associated with the grid cell size of each individual cell. With vector systems, the attributes within the spaces between and among points repre-senting the various geographic objects are assumed to be uniform unless explicitly encoded in the vector attribute tables. For example, the attributes between two links (nodes) along a vector-based road network will generally represent only a single value for each attribute. So a road's condition my be classified as "in need of repair." This condition remains the same along the entire length of the line between nodes. To show that part of the road is not in need of repair requires the user to insert another node that allows him or her to change its attributes. This limitation of the vector data model can be avoided if your software contains some form of dynamic segmentation, designed specifically to allow changes along a linear object such as a road. In raster, a road that is represented by a series of grid cells could change in attribute along its length simply by the user's assigning different grid cell values. So a road could range from "no repairs needed" to "minor repairs needed" to "major repairs needed" simply by the user's assigning changing grid cell values along its length. A value of 1 could indicate no repairs needed, 2 could show a need for minor repairs, and 3 could be used for major repairs.

The polygon is perhaps the easiest way to envision Tomlin's premise that raster models are position oriented and vector is more theme oriented. In vector GIS, the polygon is represented by an encompassing set of lines, each of which is defined by a pair of coordinates. The polygon itself has no other entities to define it. Its attrib-utes are uniformly assigned for the encompassed space. It might represent a type of land use, a class of agriculture, or a soil type. This leaves no ability to account for internal variability or for indeterminate boundaries for that matter. Raster data mod-els can include a range of values to indicate both internal variability and indetermi-nate boundaries if necessary. For example, a land use polygon in raster can be indicated by values ranging from 1 to 10 indicating intensity of that land use. The concept of geographic region is readily employed here because it allows polygons of a single land use type to have this internal variability but still remain part of a larger regional class. Soils polygons represented as raster grid cells also allow for the recog-nition that within any given soil type there may be a range of texture, pH, horizon depths, and many other values of interest to the soil scientist, agronomist, or envi-ronmental professional.

On the basis of Tomlin's idea that raster is a position-oriented tessellation, Tomlin showed that they are more useful for answering "where" questions, whereas vector tesselations are more readily adapted to answering "what" questions. Although there are individual cases in which this is not necessarily true, it is generally the case for most applications. Perhaps because of the exceptions to the general rule—the idea that raster GIS is better used for answering "where" questions, this can seem just as counterintuitive as its position orientation, but a few quick examples might prove his ideas to be quite valid. Because polygons provide the most easily understood objects for this, I will restrict my simple examples to them. Moreover, because our focus here is on raster GIS modeling, I will focus on the utility of the raster tessellation for answering some "where" questions.

Let us say that you wish to use GIS to examine the application of fertilizer to your crops to improve yields. In a vector GIS, your crop is most likely going to be represented as a single polygon. You could improve your situation by creating polygons indicating, for example, polygons of soil nutrient availability. The polygons will most often be range-graded—that is, showing a range of variability within a single polygon. A raster representation of the soil nutrient data would most often indicate a gradual changing of nutrient availability, each grid cell containing its own value. The raster representation shows much more detail about the spatial variability of nutrients within your field than does the vector counterpart. This will allow you to determine exactly where (within the field itself) the addition of larger amounts of fertilizer would likely be most effective and exactly where (within the field) less fertilizer is needed. It provides you with a wider range of possibilities. Additionally, the gradually changing nature of the grid cell representation might also prove useful for identifying trends in nutrient variability that might be associated with slope or other factors. This additional information would also prove useful for future planning of cropping and amendment applications, possibly suggesting a need for some minor terracing.

Another example of how the "where" question is more easily addressed with a raster GIS over a vector type deals with the movement of material either across the surface or within substrates. Such flow modeling is much easier to perform with raster GIS than with vector. Let us say, for example, that you are trying to examine the movement of nonpoint source pollutants, perhaps from the field we just added fertilizer to, into nearby streams. The fundamental questions being asked are where the pollutant is coming from, where it is going, and how much is getting there. Because our field is not uniform throughout, and because the substrate is also not uniform in its ability to move the dissolved fertilizer, a grid cell representation will allow us to account for this internal surface and subsurface variability.

As you might imagine from the last example, nearly any modeling requiring flows across an entire grid, whether that grid represents a surface, subsurface, or even atmospheric conditions, would most effectively be performed with a raster GIS. Such tasks as predicting the movement of harmful gaseous plumes from hazardous chemical spills, the movement of oil from tanker spills, the effects of shelterbelts on erosion reduction, the dispersion of seed rain from trees, and the movements of animals over terrain and even designing wind farms all require the ability to quantify variations in change from one place to another. The modeling of these changes requires us to know where the variability is and how that can be translated into predictions of where our modeled object will ultimately be.

There are, of course, other advantages to raster GIS modeling, many of them stemming from the increased availability of raster forms of satellite, aircraft, and ground-based remotely sensed imagery. Although there are ample algorithms to allow us to convert from raster to vector and back again, the ease with which raster remotely sensed data can be input to a raster GIS makes it an easy choice for modelers. This is particularly true when remotely sensed data are being applied to the database to update attribute changes through time. Such updating and time–change analysis techniques most often employ some form of map overlay technique. And although vector map overlay is readily available, the raster approach has some advantages over the vector equivalent. First, raster overlay is typically faster computationally than the vector approach, especially when using a simple raster data model rather than an extended model type. Second, there are a number of mathematical overlay techniques that can be applied more easily within the raster domain than in vector. Finally, the raster tessellation eliminates some of the problems of vector overlay related to the development of sliver polygons, very small polygons that may not accurately reflect the locations of attributes. In complex vector databases, when overlay operations are performed, the user must contend with what may prove to be an overwhelming number of sliver polygons resulting from the operation. The primary problem is ascertaining whether the slivers actually represent real change or are artifacts of the spatial error component of the vector data sets.

Keep in mind that although raster has some advantages over vector specifically related to map overlay, there are also some equally disturbing problems that the raster tessellation also incorporates into the process. Although two grid cells, one from each of two different layers, when compared through overlay operations, will typically yield a single value with no silver polygons to contend with, the internal accuracy of the grid cell presents its own question about data validity. Because each grid cell contains some "average" or spatially generalized value through the process of quantization on input, the results of the comparison are not always exact. In other words, though the process of raster overlay is frequently easier for the interpreter to examine, the results are not always going to be any more accurate in terms of what the resultant value is. It will, however, prove more useful for examining where the change is most likely occurring.

Among the more typically stated deficiencies of the grid cell over vector are its less attractive aesthetic appeal (from a cartographic sense) over vector GIS, the often enormous size of the databases required to store raster GIS maps, and the resultant computational expense necessary to perform operations on such large data sets. An early raster modeling research effort examined these in some detail, with particular concern for possible implementation problems related to the size of the database (Williams 1985). Rapid improvements in computer technology, particularly with the increased sizes of storage devices, improved compaction algorithms, and increased processor speeds, reduce the problems associated with both the storage and analytical speed applied to raster GIS databases. Some of these computer improvements have been somewhat minimized by the advent of floating-point data and mathematics associated with the newer raster models, as opposed to the integer data storage and analysis of older software systems. Still, for most raster GIS applications, the current technology is sufficient. These improvements also allow much larger and more realistic data sets to be developed, stored, and analyzed at the same time that the grid cell size is reduced in size. Just as an example, new remote sensing technologies will soon be readily available in 1-meter resolutions. The reduced grid cell size has the nice side effect of producing output that is far less blocky in appearance than the small data sets with large grid cell sizes common in the 1990s. In fact, many of today's fine-resolution raster data sets appear more aesthetically pleasing that their vector counterparts because of their ability to illustrate gradual changes and variation in attributes. In short, although raster databases take up more space and do require substantial computing power, especially for more advanced flow-related algorithms, the flexibility and power of their modeling capabilities far outweigh these limitations.

SOURCES OF DATA

It has been pretty much acknowledged, industrywide, that a substantial portion of the expense of running a GIS operation will come from the conversion of analog forms of spatial data and information to their digital alternative. This is certainly no less true of raster data than of vector data. Until now, we have discussed raster data in rather general terms, with the exception of digital remotely sensed data. It has been pretty much assumed that much of your data will be input from analog maps through some form of digitizing process. This is by far the slowest, most tedious approach to obtaining raster GIS. It does allow you to have greater control over the quality and applicability of the input data. If time is a problem, however, more and more quality raster data sets are becoming commonplace through local, state, and national organizations around the world. Some government organizations archive vast quantities of digital data, both raster and vector, and a whole new industry is being developed around creating and providing raw, value-added, and custom data sets for purchase. Although there are some institutional questions that must be addressed prior to obtaining and using such data sets, I refer the reader to other texts dealing with such issues, rather than addressing them here (Chrisman 1997, DeMers 2000a). Instead, I will focus on some readily available types of raster GIS data, some of their sources, and sources of custom data sets for your own applications.

As you might guess, there are many types of raster data that can be obtained, many formats, and many sources of these same data. Typical types of raster data include commercial digital remote sensing data whose grid cells are actually picture elements (pixels), scanned imagery such as digital orthophotoquads, LULC (land use/land cover), digital raster graphics (DRGs), digital elevation models (DEMs), and National Wetlands Inventory (NWI) data. Recently, a number of GIS data clearinghouses have begun developing at state and regional scales. Additionally, some government organizations have funded similar types of clearinghouses especially focusing on research applications of the data to determine its utility for a variety of real-world settings. New companies are springing up dealing either in the selling of raw or value-added GIS data or in the data exchange among a number of data-sharing institutions and individuals. Some of these private-sector companies are attempting to keep the costs of digital data to a minimum by asking users to provide data sets that they created for themselves free of charge, or in exchange for other data sets within their clearinghouse. GISDataDepot is an example of a large firm that operates within this framework and shows promise with a large and growing set of state, regional, national, and even international data. This is still a new approach to commercial data providers. Many private firms that specialize in access to digital data often charge high rates for data that are sometimes available for much less when not packaged as value added. Federal, provincial, and state government agencies are beginning to join forces to collect base data for their potential clients. In the United States, the U.S. Geological Survey, through its National Spatial Data Infrastructure (NSDI), is currently developing a National Geospatial Data Clearinghouse. The idea for this system is to create a set of distributed sites organized under the four themes of biological resources information, geologic information, national mapping information, and water resources information. This is resulting in a set of state-based clearinghouses that serve both the individual states and the nation. Similar programs are also developing in Canada. For example, the provincial governments of Alberta and British Columbia have defined and implemented the development of provincewide data sets for forest inventory.

Government-provided data are often available for large regions of the world, at a reasonable cost, and are often an extremely good source of base map data (Kemp 1993). However, the reduced cost and large area coverage is sometimes offset by reduction in the data's quality, temporal usefulness, and accuracy. Availability of government-provided data sets varies, with general availability and large area coverage most often available at the federal level. In the United States, a good place to start is the Manual of Federal Geographic Data Products, much of which is available on the World Wide Web, from the Federal Geographic Data Committee (FGDC) (1992). This will provide a source of data from many U.S. federal agencies such as the Bureau of the Census, Natural Resources Conservation Service, National Aeronautics and Space

Administration (NASA), the Federal Emergency Management Agency, Bureau of Land Management, and others already mentioned. The U.S. Geological Survey Publications and Data Products Page also provides an index of available data products, including some already listed by the FGDC. The FGDC also provides a large suite of automated products for recording your metadata so that future users will have a clear understanding of the nature, quality, validity, and lineage of the data you are using in your work. We will discuss metadata a little later on when we examine error in GIS.

Government data available at the regional, state, or local levels present some unique problems for acquisition. In many cases, the data are created, stored, and disseminated in formats that are not immediately compatible with your own needs. Additionally, their availability may vary on the basis of the agencies' own established procedures for dissemination, the sensitivity of their data, and even the willingness of agency personnel to provide the data. In many cases, it will be difficult for you to even define where, within a government body, a particular type of data may reside. Getting data from these agencies may require developing working relationships with the personnel themselves. This may involve a substantial investment in time, but such an investment frequently proves beneficial to both yourself and the data providers, who may share similar modeling needs.

Commercial sources of data are becoming more commonplace as the need for geospatial data grows. These providers may have access to value-added governmental data, data sets developed by their own clients, or data developed in-house for their own projects. Depending on the nature of the company and its relationship with other clients, its representatives may be willing to provide you with data directly or to provide contacts with their clients for obtaining data sets. In many cases, data providers will, for the appropriate price, be able to provide data conversion services for converting your analog data into compatible forms of digital data. One efficient method of locating data providers is to contact your GIS software providers. They frequently know who the major data providers are and often have working professional relationships with them as well.

Although using available digital datasets has the obvious advantages of not requiring the time and expense of data conversion, there are some important things you need to keep in mind when deciding to use them. First, just because the data are available does not mean that you need to use them. If they are not the exact data you need for modeling, don't use them. Second, although some data are quite compatible with your application, many more are not. For example, the resolution, projection, study area, classifications, and other data characteristics may differ radically from what you might have envisioned for your modeling needs. Don't let the data drive your model. Third, even if the data seem to fit your needs in terms of their characteristics, it is important that you know the lineage and quality of the data. If you have no quality control, you compromise the model results as well. Some data sets come with little documentation. Today you should expect to have some form of data quality assessment, together with a complete description of what the files contain, how they were compiled, their sources, and what methodologies were applied to their quality assessment. For extensive data sets, such a report is most often provided as a separate publication, often called metadata, that includes a detailed data dictionary. Software is available from a variety of sources on the World Wide Web that provide fairly good methods of compiling the metadata, and in such a manner that they comply with the Federal Geographic Data Standards Committee. If the data you are obtaining does not have complete documentation, you may be forced to create your own from analog data sources.

This, of course, suggests that your need to perform the data conversion yourself is primarily to maintain quality control. Another pertinent reason for such conversion is the widely varied data formats, each possibly requiring converting from one digital format to another, presenting the possibility of introducing unexpected data error. But converting your own data has its own risks as well, particularly the substantial time it takes to perform the conversion and the substantial costs associated with it. A prototype conversion from analog to digital of a portion of the relevant data types for your project may provide information about the amount of time required, the attendant costs, and the reliability of the software to produce results of a quality superior to what is already available in digital form. Ultimately, the decision to develop your own digital database should be made with a complete knowledge of the necessary data for your project, the digital alternatives, and a complete evaluation of your own conversion capabilities. Of course, should you decide to develop your own database, you should also consider that you must also develop a detailed set of metadata for your own internal use. This is especially important if the data are to be used for long-term projects or if they will be sold as data sets to other GIS users.

SELECTING PROPERTIES: GRID SIZE, STUDY AREA, DATA FORMAT, PROJECTION, AND GRID SYSTEM

Modern GIS software is becoming quite sophisticated. No longer restricted to the Cartesian coordinate system, these new GIS packages allow for the input and interaction of a wide range of grid systems and projections. Additionally, as we have already seen, the advent of faster computer processors and massive storage devices has reduced the need for an inordinate concern over all but the largest databases. Still, these decisions will impact the availability of the GIS data set, its accuracy, your ability to model with it, and the resulting quality of output.

Among the primary decisions to be made in developing a raster GIS database is the grid cell resolution. Although in the past such decisions were largely related to the size of the database and the ability of the software to handle large data sets (Williams 1985), today the cell resolution is more closely linked to the modeling needs. So what dictates the size of grid cell selected for a given modeling project? If grid cells were one-dimensional, sampling theory (Shannon and Weaver 1949) would dictate that the size of grid cell should be at most one half the size of the smallest object needing to be mapped (the minimum mapping unit). But grid cells are two-dimensional. Therefore, the grid cells should be at most one fourth the size of the minimum mapping unit. In these circumstances, this would provide for at least four grid cells for each object to be mapped (Figure 2.12). This general rule is easily understood and seems perfectly reasonable. If the objects being mapped are long, sinuous polygons, for example, such a simplistic sampling strategy can result in necessary objects' being omitted from the database (Figure 2.13). This suggests that care be taken in using sampling theory to define the grid cell size.

There are other methodologies that can be applied to grid cell selection. One common one is the matching of grid cell sizes to the pixel sizes of digital remotely sensed data that are to be employed within the model. Generally, it is not a good idea to let the pixel size of remotely sensed data limit the quality of your model, but if much of your model revolves around those data, and if the pixel resolution is otherwise

Figure 2.12 Sampling your objects with grid cells. At least four grid cells should be allowed to represent the smallest object with which your geographic information system model must perform its analysis.

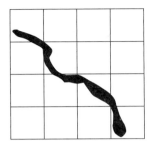

Figure 2.13 **Although simple methods of deciding how many grid cells should be used to represent the smallest mapping unit work for compact objects, long, sinuous objects may require more than four grid cells to represent them.**

acceptable for that model, it is a practical and expedient solution. The sensitivity of the model to grid cell size is a necessary first step in this case and may suggest a simple prototype examination of how the model functions under different grid cell sizes (DeMers 1992). A unique approach to selecting the grid cell size for raster GIS modeling, and one that considers the actual scales at which the data in the model interact with one another, uses hierarchy theory (King et al. 1991). Such approaches require a detailed knowledge of the nature of the data and the model itself. Something that should be considered no matter what method is used is grid cell size.

Related to the grid cell size is the size of the study area. Even with the most powerful computers, you would not want to use a 1-meter grid cell size if you are trying to model Australia. Of more interest is the relationship between the areal extent of the study area and the types of modeling functions to be used. This is particularly important when any form of interpolation is used. In such a case, it is recommended that the study area actually extend beyond the actual land area to be examined. This is to ensure that the interpolation algorithm has ample data points to perform its operations. Once complete, it can be clipped and incorporated into the remainder of the database (DeMers 2000a). The primary exception to this approach is when the outside limit of the study area is discrete and absolute, as in the case of an island.

Data formats are becoming increasingly compatible with one another as data providers, software vendors, and users all begin to recognize the utility of compatible data sets. Raster data sets are no different than vector data sets in that they come in different formats, each of which is more or less compatible with other types. These types range from image formats from scanning operations (e.g., JPEG and GIF files, as well as satellite remotely sensed data), to those created by digitizing operations, to a variety of raster GIS formats including both simple raster and extended raster, as well as data that have been converted from vector to raster. The implementation of the spatial data transfer standards (SDTS) has provided some order to data sets whose origins are funded by U.S. federal funds, but there are still many varied types and sources. Rather than try to list all available data types, which is outside the intent of this book, I simply leave you with this caution: Know the data types that your particular software supports, in particular those that it is capable of importing and using. These are most often provided in your GIS software documentation and should be referred to prior to selecting data to be used for your project. It is also worthwhile to consider the ability of your GIS software to export data into other formats. There are often cases in which a particular piece of software that you would not normally use for modeling may be ideal for an especially complex or unique task. Should you need to export selected grid themes to this software, you will need to know its data requirements for compatibility. This most often occurs in GIS operations in which a superset of available GIS and remote sensing packages are used, especially where contract work or other consulting activities require multiple platforms, software systems, and data types.

The same can be said for grid systems and map projections as for data types. Although GIS and remote sensing software are capable of converting from one map projection to another and from one grid system to another, such operations generally introduce unnecessary error into your GIS database through computer rounding error during computation of these new configurations. Burrough and McDonnell (1998) clearly demonstrated that the more often these operations take place, the greater is the potential for otherwise small errors to compound to several orders of magnitude of difference. Most GIS software performs its analysis under a single map projection (Wang 1999) or converts the data to geographic coordinates for analysis. If your database contains multiple projections and multiple grid systems, these will, of necessity, be converted for analysis. The fewer conversions, the less the chance of computational errors creeping into your models.

DEALING WITH THE ERROR COMPONENT IN RASTER DATA

No GIS database is error free. It doesn't matter if the database is vector or raster. Because maps are essentially models of spatial reality, it is impossible to have an analog map or a digital map that does not contain some level of uncertainty or error. This error can be bothersome in cartography, but when applied to the analysis of cartographic data, it becomes exasperating because of the ways in which the errors permeate the model itself. Raster GIS models often assume that the source data have little error, that the input methods introduce little additional error, that all the boundaries are discrete and easily defined, that class intervals and categories are equally useful for different applications, that the algorithms that link these attributes are totally deterministic, and that the results of algorithmic manipulations will be equally deterministic. Since the 1980s, there has been a huge volume of research performed on identifying, characterizing, quantifying, eliminating, minimizing, and otherwise dealing with the error component of the GIS. The volume of literature devoted to the subject even suggests predilection for the topic. This focus on what has turned out to be a really nasty, highly complex problem is due to an industrywide acknowledgment of the potential problems of working with erroneous databases, the difficulties of actually implementing models using incomplete or otherwise faulty data, and the financial and/or litigation impacts of poor decision making as a result of incorrect model results. Error can be considered for each individual thematic map or grid that is included in the GIS database, it can include the interaction of these less-than-perfect grids throughout the modeling process, or it can focus on the quality of the final output of the modeling process. We will examine GIS data, strictly within the raster environment, and from three different perspectives—data representation (including input), error propagation within the modeling environment, and methods of identifying and coping with error from both a data representational standpoint and from the error propagation perspective.

Although we have enumerated the implicit assumptions that either directly or indirectly contribute to modeling difficulties with erroneous data, it is important to identify the typical sources of raster data error. These are typically the same for raster as they are for vector and include accuracy of content, measurement error, field data collection error, laboratory error, locational error, and error due to natural spatial variation (Burrough and McDonnell 1998). These types of error commonly occur prior to or at the input phase of the GIS project. Content error includes both qualitative types stemming from misclassifications or misidentifications of the categorical content of the grid cells at the nominal scale of data measurement, and quantitative error is more typically caused by bias or improperly calibrated data collection devices such as rain gauges, pH meters, and telemetry devices. Categorical error is common in both vector and raster systems, but the raster tessellation compounds the error, especially where a single category is assigned to each grid cell. Simple comparisons of the total amount or percentage amount of categories found in the analog document and the raster equivalent shows the impact of the raster tessellation on the amounts of each category. Employing a ratio of each category in the analog input map to the raster equivalent demonstrates the loss or gain of information through the input process. Quantitative content error is closely allied to measurement, field data collection, and laboratory collection errors prior to GIS input and suggest that care be taken in categorizing data and calibrating instruments prior to data collection. Locational accuracy deals explicitly with the accuracy (closeness to actual values) and precision (how closely multiple measurements provide the same information) of the surveying instrumentation used to determine absolute locations on the earth's surface. Two typical problems associated with locational accuracy stem from survey error for field surveys and from pixel displacement while employing digital remotely sensed data. Modern global positioning system (GPS) equipment provides us with accuracies that are typically well within the needs of most raster GIS models and are very useful in locating individual coordinates for remote sensing pixels. However, locational accuracy can also occur during the input phase of the GIS operation where digitizing analog maps is involved, especially when unstable map documents are subjected to stretching and shrinking caused by changes in temperature and humidity levels (DeMers 2000a, 2000b). This suggests that maps made of stable materials such as Mylar, as opposed to paper, should be employed wherever possible or that temperature and humidity be tightly controlled when they are not available.

Among the most difficult and perplexing errors are those resulting from GIS input of naturally varying components of the earth. Vegetation, elevation, and soils are common types of earth components that contain both local variation that cannot be accounted for at a given scale and whose categories change as a continuum. As with all continuous surfaces such as topography, with many highly varying discrete surfaces we are forced to sample the data to provide synoptic coverage. In remote sensing, for example, we limit our input to the size of the pixel observable with the sensor. When our surface varies substantially within the available pixel size, the sensor averages these values, resulting in mixed pixels. Such accuracy problems due to natural variations are unavoidable, but their impact can be lessened by selecting grid cell or pixel sizes that are more likely to contain relatively homogeneous categories and by adding ancillary data to account for indeterminate boundaries. A rather comprehensive and well-thought-out analysis of some more exact methods of dealing with such errors can be found in Burrough and McDonnell (1998). Their text also clearly defines the role of data age, areal coverage, map scale and resolution, observation density, relevance, data format, accessibility, costs and copyright, and computational errors that affect the reliability of spatial data inside a GIS. These subjects are very important to the quality of our modeling but are beyond the scope of this text.

Given our focus on modeling, however, it is vital that we examine the problem of error propagation within our GIS modeling process. Given that we have less-than-perfect data provided as input to our GIS, it might be concluded that if we can limit that error, our problems associated with error are behind us. Instead, we find just the opposite. When modeling in a GIS, we have three primary factors of error propagation that cause potential problems with model output. These are the quality of the data, as we have already seen, the quality of the model itself, and the interactions of the data with the model. An examination of error propagation in GIS modeling requires estimates of errors in the source data, error propagation theory, and error propagation tools (Burrough and McDonnell 1998).

Error estimates for input grid themes most often involve some form of stochastic simulation because few geographic processes, whether human or natural, are so clearly understood that deterministic models of error could be employed. The primary assumption for stochastic simulations is that the data are normally distributed. This is clearly not the case for many data types, but in the absence of a clear understanding of the underlying processes, such distributional patterns are employed to produce error surfaces. This is most commonly done through a general study of error propagation itself. A brute-force method of examining the error in a GIS model is called a Monte Carlo simulation; it assumes that each attribute has a normal distribution. In this case, if we add a new attribute, U, as a function of inputs A_1, A_2, ... A_n, we want to examine the error associated with U and what the contributions of each of the functional inputs A_n are to that error. Successful examples of the use of Monte Carlo simulation error analysis have been used to examine numerical surface flow models (Desmet 1997) and soil map inclusions (Fisher 1991). The Monte Carlo simulation method requires dozens, often hundreds, of simulations to be run, depending on the size of the database. It is useful when the grid cells interact spatially within the model through functions like neighborhood and windowing operations, interpolation processes, and buffering.

Although it is relatively straightforward, a major problem with the Monte Carlo simulation approach to error propagation modeling is that it is a brute-force method, requiring a substantial investment of computing resources. These often limit their usefulness in typical GIS modeling efforts, where tight deadlines and already overtaxed computing resources suggest that the modeling itself takes presidence over an estimation of error propagation. An alternative approach is to use standard statistical theory of error propagation, suggested by Parrat (1961) and Taylor (1982). This approach, generally known as point analysis, is useful when the model employs local functions (see Chapter 4), where the grid cells do not interact with each other spatially. Models built primarily using typical overlay operations would be good categories for such error analysis. The basic premise of point analysis is that for each grid cell there is an error component that is a unique function of the input values when the transformations are limited to arithmetic relationships. For such an error analysis, there is a computer program called ADAM that can trace such point mode error relationships (Heuvelink et al. 1993, Heuvelink et al. 1989).

Both of these techniques require a detailed knowledge of your database, its thematically unique error components, and, in the case of point analysis, limitations on the types of model employed. There is still much that we do not know both about how natural processes affect the error component of each grid theme we employ and about how these interact in complex models. Given the typical time constraints, and nonresearch agenda of most GIS models, it is not very pragmatic to expect that detailed error analysis and pedigree development be performed on every model, or even on prototypes of these models. Still, to present model results without some form of verification and validation could prove dangerous. Among the best solutions is to perform a model verification through the employment of a truth set against which the model results can be compared. This is more than simply suggesting that the model "looks right," a method often employed to suggest that it produced the correct result. Instead, a small sample of the study area is examined through means other than GIS to determine the correct output. We will examine this in more detail in Chapter 9.

TEMPORALITY IN SPATIAL DATA

Although research is still underway that aims at creating explicit data models designed exclusively for modeling spatial dynamics, there has not as yet been a clear, fully operational solution. There are, however, some solutions that allow us to use existing raster models for dynamic modeling. There seem to be three typical solutions to this. All of these use some form of stepwise or discrete difference modeling approach that proceeds step-by-step, where each new step is the result of some GIS function applied to a new set of grid cells but impacted by the output results and conditions resulting from the previous step.

The most basic approach to temporal modeling in GIS might be called implicitly temporal and employs the global functional capabilities (see Chapter 4) of the raster GIS. This approach is essentially a measure of distance from some starting point outward and measures the distance outward on a cell-by-cell basis. The distance can be modified by friction values embedded in another grid theme that results in a slower movement across the entire grid.

In the second type of spatiotemporal model, other conditions, most often probabilistic or logically conditional, can also be embedded in the thematic grid cells so that for each iteration of the model, the impact of those conditions can be evaluated and the results can be output to an interim grid. This has the impact of changing the conditions for each step in the model. A classic example is a fire model in which a forest grid cell adjacent to another grid cell that is already on fire will not necessarily immediately burn as the next time step occurs (Yuan, 1994, 1997). Instead, its condition will be a function of how dry it will have become during each successive step. As the adjacent grid cell continues to dry, its susceptibility to fire increases until it reaches a certain threshold value, at which time the value of the cell changes to indicate that the trees within that grid cell are on fire. Although the immediately adjacent grid cells go through this conditional transformation, grid cells at a distance from our burning cells can also change accordingly, perhaps on the basis of some distance decay function. Additionally, the raster GIS model allows for transition rules to be examined through random chance by employing the Monte Carlo simulation as a spatiotemporal simulation model. This approach has been employed sucessfully for simulating residential location choice, where the target cells are less readily identified than they might be in fire modeling (Raju et al. 1998).

We have also seen that specific raster models called cellular automata are specifically designed for incorporating conditional data for growth models (Batty and Xie 1994). And we have seen how they differ in their implementation. Because they employ only adjacent neighborhood cells, and because they can contain an explicit set of rules, often complex rules, they are considered a third type of spatiotemporal model. The fundamental difference between the cellular automata model and the raster GIS is in the advanced capabilities of the raster GIS based on map algebra. Cellular automata can perform a wide variety of complex spatiotemporal modeling tasks, but they are limited in their ability to perform overlay operations, distance measurements, and a host of extended neighborhood operations. This explains the attempts to link raster GIS with cellular automata rather than to be forced to choose between them.

Some research has also been employed to explicitly link software such as Stella, which is designed for temporal input–output modeling tasks, with the spatial capabilities of the GIS. The earlier attempts essentially performed the temporal modeling within the temporal modeling package and simply passed each iteration's results to the GIS for display. By creating regions with identical conditions, the software could operate on the regions themselves through time, rather than on each individual grid cell. More recently, attempts have been made to explicitly link the two disparate software systems for explicit space–time modeling. The results will be a hybrid system, with possibly new data structures and probably new data models. Such attempts are worth noting, but, given the experimental nature of this work, we will not examine it in detail. Instead, read some of the relevant literature regarding this new technology.

Chapter Review

Raster geographic information systems (GISs) are based on a cellular tessellation of space where the geographic space is separated into unique packets of data. There are at least four basic data models based on this tessellation that are commonly applied to modeling tasks: simple raster, extended raster, quadtrees, and cellular automata. These models are most often considered to be best applied to position-based modeling tasks, as opposed to vector models that are more thematically based. Although raster tessellations provide less accurate absolute locations, they provide the best opportunities for modeling any type of surface, and for examining the spatial interactions of phenomena, whether they are adjacent or some distance from the target cell.

There are many sources of digital raster data, ranging from free or inexpensive government-created data sets to more costly commercial data sets. Most data suppliers also provide custom data services specific to your own project. Although in-house database development allows more local control of data quality, the reduced cost of existing data sets may offset this factor as long as the data suppliers provide detailed metadata. In the data selection process, decisions concerning grid cell size, study area, data format, map projection, and grid system are necessary to modeling. Grid cell size should be linked to modeling needs, rather than to either data volume or pixel sizes of remotely sensed data. When one is making decisions about study area, it is important to extend the study area beyond the absolute area under consideration wherever surface modeling and/or interpolation is to be employed. Although data formats, map projection, and grid systems can all be converted, it is best to keep the number of transformations to a minimum to reduce the introduction of computational error into the database.

The sources of raster data error are essentially the same as for vector data—accuracy of content, measurement error, field data collection error, laboratory error, locational error, and error due to natural spatial variation. These error types are important in themselves, but their interaction during modeling is even more important given the nature of the GIS as a modeling tool. There are two primary methods of modeling error propagation in a GIS. The first is a brute-force method using a Monte Carlo simulation, which, although computationally expensive, allows error to be traced for models that interact spatially. For models that are limited to local functions or functions that do not employ spatial interaction, there is a less computationally expensive method: point analysis. Beyond these approaches to error analysis is the use of truth sets for model verification that requires a priori knowledge of what the expected results of the model are.

Although no spatiotemporal raster data model as yet exists, the existing models allow for space–time modeling. Three general types of space–time models are capable with existing raster data models. They include a simple distance measure approach where each step in the model corresponds to a time step, a second employs conditional responses and thresholding, and a third employs the cellular automata to embed the conditions within its grid cells. Research is underway to both create spatiotemporal data models and link temporal modeling software with a GIS.

Discussion Topics

1. Why do you suppose that with most raster geographic information system (GIS) tessellations and data models, the cell values are most often encoded as whole numbers rather than rational numbers? What technical advantages and disadvantages might there be with encoding them as rational numbers? What modeling advantages and disadvantages might there be?

2. Although the raster tessellation is typically done with square grid cells, consider what advantages there might be for using some alternative data structures for modeling one or more of the following:

a. Some or all of spherical bodies such as moons or whole planets

b. Small planetary bodies with highly irregular shapes, such as asteroids

c. Geological formations such as volcanic necks, plug domes, and salt domes

d. Subsurface ore bodies in full three dimensions

3. Discuss the advantages and disadvantages of using the different types of raster GIS models, including the variants of the simple raster model.

4. You have seen how the extended raster GIS model can be used for point, line, and area features. Why would this model be less effective for surfaces and fields? Include in your discussion the idea of rational numbers versus integers.

5. Explain the difference between raster GIS models (including quadtrees) and cellular automata. Why do you suppose there has been so little communication between those who model with cellular automata and those who use raster GIS? What are some of the issues involved in making those linkages?

6. Discuss the potential for advancements in raster GIS data storage and analysis in terms of what one might expect from computer technology changes. For example, track the general sizes of hard disk drives, as well as processors and their clock speeds over the past 5 years. Chart these changes using either a line graph or a histogram. Use these as surrogates for the sizes of databases that could be employed and for the speed of the analytical operations. What other factors might come into play?

7. Discuss why raster GIS is considered to be location based whereas vector GIS is more theme based. Relate this concept to the idea that raster GIS is better at answering "where" questions whereas vector is better at answering "what" questions. Give an example of this not found in your text.

8. Describe in your own words the components of raster data that are likely to cause error in your modeling activities. How can you minimize both the sources of error and the amount of error in your database? What techniques are available for modeling raster error propagation?

9. Discuss what properties might be required to create a raster data model that explicitly models spatiotemporal conditions.

Learning Activities

1. Illustrate how you would represent each of the following using a simple form of the raster tessellation discussed in this chapter:

a. Telephone poles

b. Roads and streets

c. Agriculture

d. Topography

2. Using the diagrams in your text, explain why the MAP (Map Analysis Package) raster data model is the most appropriate for extending the raster data model with the use of relational database management systems.

3. Create a table similar to Table 2.1 showing what types of additional attribute data could be represented within the extended raster data model for the following thematic coverage types:

    a. Land use

    b. Transportation net

    c. Urban infrastructure

4. Draw a simple 4″ × 4″ map showing only land (shaded) and water (white) on a sheet of paper. Create a 4″ × 4″ grid with divisions of $1/4$ inch. Using a photocopier, make a clear acetate overlay of the grid. Now, overlay the grid on the map you produced. Create a diagram showing the application of quadtrees to represent this map.

5. Get on the Internet and search for sources of raster GIS data sets. Be sure to search both government and commercial sources. Also look for sources of data beyond your own national boundaries. Store all promising links as bookmarks in your Web browser and include some copies of useful Web pages in the notebook you started in Chapter 1.

6. Using the 4″ × 4″ map and the associated grid you created in Learning Activity 4, overlay the grid on the map and simulate the dominant method of digitizing by storing your values in a matrix of numbers (16 × 16). Examine the output map and the input map. Now, assuming that each grid square (each $1/4$ inch square) is 1 square mile, determine the number of square miles of land versus water in the new map. Using a planimeter, determine determine the number of square miles of land versus water on the output map. Describe and quantify the amount of error produced through your coding process. Where does most of the error seem to be showing up? How could you determine if reducing the grid cell size to $1/8$ inch per square would improve the quality of the output map?

7. Using a 16 × 16 grid on a piece of paper, show how you could use a form of spread function (distance function) to simulate the movement of something from the upper left-hand corner to every other grid cell throughout the rest of the grid. Now add your own friction surface and demonstrate the stepwise procedure again. Now speculate on how you could incorporate thresholds and logical rules to simulate the spread of something like fire. We will revisit this in more detail later, so don't spend an inordinate amount of time on it.

## CHAPTER 3

Map Algebra

On completing this chapter and combining its contents with outside readings, research, and hands-on experiences, the student should be able to do the following:

1. Identify and implement appropriate coding schemes for point, line, area, and surface data at all levels of geographic data measurement

2. Explain the difference between systematic and nonsystematic raster encoding and discuss the advantages and disadvantages of each

3. Create extended raster attribute tables for point, line, area, and surface data at all levels of geographic data measurement

4. Analyze and quantify single-theme spatial error created through one or more encoding schemes for raster data

5. Describe and illustrate the methods of dealing with encoding multiple point and line entities and their associated attributes when multiple objects occur within the geographic boundaries of a single grid cell

6. Explain the similarities and differences between map algebra and matrix algebra

7. List and define the basic operators available in map algebra and provide brief descriptions of each, including suggestions as to what they would be used for

8. List and define each of the functions of map algebra, provide brief descriptions of their purpose, and suggest, in brief, how one might employ them

9. List and define each of the basic flow control operations available in map algebra and explain why flow control operations are necessary and how they might impact the automation of modeling operations and functions

10. Explain what iteration operations are and what role they might play in the development of cartographic models

CONCEPTUALIZING ZERO- THROUGH TWO-DIMENSIONAL SPACE WITH GRID CELLS

In Chapter 2, we learned about some spatial tessellations used to quantize geographic space into discrete units or grids. We also learned some basic methods for modeling multiple thematic data within the computer. I purposely avoided a discussion of data measurement levels until now, when we combine our tessellations, raster data models, and mathematics within the central framework of cartographic modeling—map algebra (Tomlin and Berry 1979). Before we can effectively discuss map algebra, we need to review the ideas of nominal, ordinal, interval, and ratio scales of geographic data measurement as outlined in most introductory GIS textbooks, but focusing explicitly on the raster tessellation of geographic space. We will begin with zero- through two-dimensional space because most coverages inside raster GIS are not statistical surfaces. Additionally, we will discuss the conversion of analog spatial data into the basic raster tessellation so that we have a feel for exactly what it is we are modeling, at what measurement level, and with what possible positional errors we will have to contend as we build our models.

Two-dimensional space will be defined here as any nonstatistical surface-related geographic data and will include the three basic cartographic data types we saw in Chapter 2—namely, points, lines, and areas. Each of these objects represents real-world features that have been characterized by the observer and then abstracted to some degree both spatially and numerically. Such point features as electric power poles actually resemble points in reality, thus not occupying any substantial area on the ground. As such, their absolute location is often considered to be somewhere within a given grid cell, and the accuracy of the location is a direct function of the grid cell size. As we have just described it, the power pole is identified only as a single piece of nominal data in that it simply has a name associated with the location. Under such circumstances, the power pole is often recorded as a numerical value (usually a whole number) that merely indicates its existence somewhere within the grid cell. To encode this, we preselect a number to represent the grid cell and note its location using a presence/absence methodology of raster encoding (DeMers 2000a) (Figure 3.1). This method typically uses some nonzero value to indicate presence of the object and zeros to show its absence. At this point, if we are using the extended raster model, we can also indicate any additional attributes, whether numerical or categorical, as part of the linked relational database management system (RDBMS) (Table 3.1). Such attributes might include the size of the pole, its type (wood, metal, etc.), and the last time it was inspected.

Let us assume, however that we would rather encode the poles not as a single category of electrical poles but as multiple categories. For example, we might want to create a theme called *power poles* that explicitly categorizes them as, for example, with one, two, and four crosspieces, or each category of pole could be coded as a

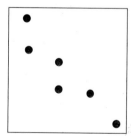

Figure 3.1 **Presence/absence coding method.** This method of encoding raster data is the best method applied to point data, although it is not limited to points. A grid cell is encoded 1 if the object is present somewhere within the grid cell, 0 if it is not.

1	0	0	0
2	2	0	0
0	3	1	0
0	0	0	1

TABLE 3.1   Extended Raster Model

Value	Count	Size	Type	Inspection
1	3	18″	Wood	1/20/97
2	2	22″	Metal	9/19/99
3	1	24″	Metal	9/30/99

Presence/absence phenomena are easier to handle with the extended raster model that allows for the inclusion of additional attributes. In this case, we are looking at power poles, each of which is of a different size, type, and inspection date.

separate theme. In this way, the original theme contains a preselected category of nominal data. Then for each, we still have the ability to store, isolate, and retrieve additional attribute data as tables in the database extension for each of the specific pole types. This has some advantages by simplifying the data and allowing ease of search for, say, poles with a single crosspiece but that have not been visited for maintenance for over 6 months. If your raster GIS does not have a direct link to a database management system (DBMS), it is probably wise to produce themes with multiple categories rather than preselect each category. In this way, the grid cells themselves contain more information because of the categories, thus reducing the need for the extended DBMS.

Of course, point objects can also be coded on the basis of ordinal, interval, and ratio categories. As before, we can encode each pole either simply as a pole, with its ordinal, interval, and ratio categories stored in an extended database as attributes, or we can explicitly encode them as numbers for simple raster databases. And as before, they can be encoded using the presence/absence methodology. Some examples are shown in Table 3.2. Note the differences between the simple and extended raster models.

We have seen that there are many options as to how these point features are selected for encoding. The systematic coding strategy we used was the presence/absence method. However, we are not limited to this method for such point objects. Although the other three systematic methods—centroid of cell, percent occurrence, and dominant type (DeMers 2000a) are not particularly useful for point objects, there is one nonsystematic raster coding method that can be used. This approach, called the most important type method (Environmental Systems Research Institute Staff 1994), allows the user to selectively isolate types of objects to be included while omitting others. This can be exemplified by selecting and encoding only those power poles that are in need of inspection in a given theme. Of course, any additional thematic data can be included in tables within the extended raster model. The most important

1	0	0	0
2	2	0	0
0	3	1	0
0	0	0	1

TABLE 3.2   Theme: Power Poles

Value	Count	Size	Type	Inspection	Crosses
1	3	18″	wood	1/20/97	0
2	2	22″	metal	9/19/99	1
3	1	24″	metal	3/30/99	2

The presence/absence method of encoding can be extended with tabular data to include multiple descriptors.

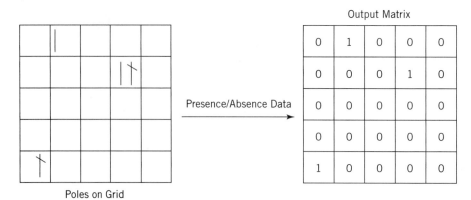

Output Matrix

0	1	0	0	0
0	0	0	1	0
0	0	0	0	0
0	0	0	0	0
1	0	0	0	0

Poles on Grid

Figure 3.2 **One problem of presence/absence encoding.** There are often situations in which two or more objects can be contained within a single grid cell, but without using an extended raster model, there is no way to note this.

type method gives the user much more control over what is important in the thematic data before modeling begins.

As we've seen, the point object has no real spatial dimension, yet the grid cell does. This creates some potential problems for situations in which two or more point objects occur within the same geographic extent of a single grid cell. For example, we may have two or three power poles that happen to occur within the extent of a grid cell (Figure 3.2). If you include a single value for each pole and you have three poles within this grid cell, you are limited to encoding only one of them with a simple raster data model. Simplifying your theme so that you can isolate poles of a particular type may eliminate this problem if each of the three poles is separable by type. Another alternative is to select a smaller grid cell size so that each pole would occur within a separate cell. Yet another approach is to encode each grid cell as either having or not having poles, then using numerical values to indicate the number of poles included for each grid cell. This approach does limit the utility of the extended raster model because you may have more than one type or category of pole at each grid cell. The extended raster model offers a similar approach to our last one in that each grid cell could be encoded as to the presence or absence of telephone poles and then the RDBMS could contain specific information about how many poles were there. As with the previous method, however, it limits the addition of attribute information. Because RDBMSs usually assume a single value for each cell (first normal form [DeMers 2000a]), you would be limited to attributes for only one pole at a time. This can be avoided by ignoring the first normal form and including separate columns for each of the three poles (Figure 3.3). Doing so will make querying the database more difficult and will limit the utility of your database. In the example from Figure 3.3, you would have to search for the value of 1 to indicate that there was a pole to begin with and then continue to search the other pole descriptors as well. This problem is well illustrated by the Mount Desert Island, Maine, database used in *Exercises in GIS* (DeMers 2000b).

Line features share the same levels of geographic data measurement (i.e., nominal, ordinal, interval, and ratio) as well as the same encoding options. Because line data have only one dimension, length, raster data structures generalize the spatial position of these features (Figure 3.4). In this way, a road or path with a width of 15 meters, for example, would occur somewhere within a grid cell that might be 100 meters in diameter. This again raises the need to be careful when selecting the grid cell size to encode line data.

Selection of an appropriate raster encoding scheme for linear data presents much the same set of problems as for point data. Because line data typically occupy a very

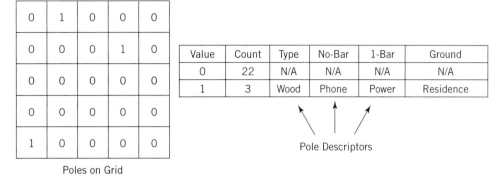

0	1	0	0	0
0	0	0	1	0
0	0	0	0	0
0	0	0	0	0
1	0	0	0	0

Poles on Grid

Value	Count	Type	No-Bar	1-Bar	Ground
0	22	N/A	N/A	N/A	N/A
1	3	Wood	Phone	Power	Residence

Pole Descriptors

Figure 3.3 Ignoring first normal form. Although putting in separate descriptors for objects may allow us to store them, it often creates problems when trying to retrieve them.

small portion of the grid cells, the use of percent occurrence, dominant type, and centroid of cell methods of raster data encoding is not appropriate. We are left with presence/absence and most important type methods. As with the point data examples, presence/absence would most often be used when the categories are kept Boolean (e.g., either we have roads or we don't). To include more or different categories or varieties of line data, we could use the most important type method. Whichever of these methods we select, we still run the risk of having more than one linear feature appearing within the limits of a single grid cell. In fact, because roads, paths, and railroads frequently cross, the possibility is even higher than that for point data. Careful selection of line data categories allows us to avoid this problem in most situations (Figure 3.5). Of course, we still have the same options, as illustrated by point data with this same problem.

Linear data can be categorical, such as roads versus paths versus railroads, each of which is normally coded within a particular theme (Figure 3.6). The previous examples would most likely be included in a theme called transportation. In this way, each category could be included within a single theme, which would be a relatively compact way of storing them. Line data can also be ordinal (e.g., single-, double-, or multiple-lane highways), interval, or ratio (e.g., based on traffic flow or measured width). Each of these could be recorded as a separate theme, or they could be included as a single theme (e.g., roads), just as in the case of nominal categories (Figure 3.7). For extended raster data models, we also have the opportunity to store additional nominal, ordinal, interval, and ratio attribute data when necessary. And, as with point data, we could encode the roads as a single category, *roads,* and then

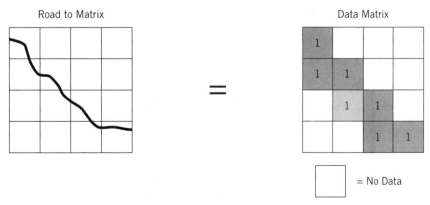

Road to Matrix

Data Matrix

= No Data

=

Figure 3.4 Generalization of lines in raster. The use of grid cells for line objects presents the problem of positional inaccuracy due to the size of the grid cell in geographic space.

Most important type method

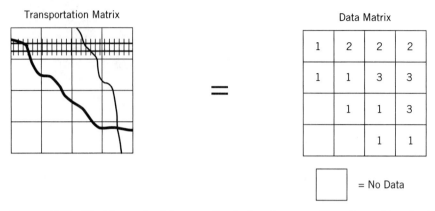

Transportation Matrix

=

Data Matrix

1	2	2	2
1	1	3	3
	1	1	3
		1	1

☐ = No Data

Figure 3.5 Most important type method of raster encoding. By making decisions about what objects are most important prior to raster encoding, you eliminate the problem of having two or more objects intersecting at the same grid cell.

add any additional attributes, such as *traffic accident reports* and *road condition*, to our tables. In this case, if, as in the case of multiple points occurring within a single grid cell, we have several road types crossing a single grid cell, we run the risk of not being able to encode all the attribute data without violating the first normal form (Figure 3.8).

Area features have two dimensions and the same options for level of geographic data measurement for their associated attributes. The two-dimensional patterns of area features offer some interesting options for encoding not available to point and

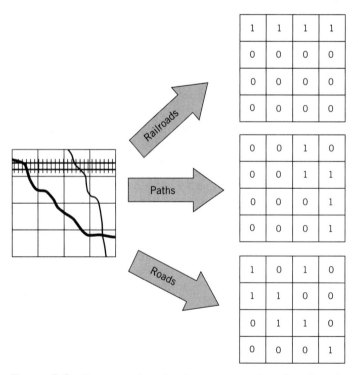

Figure 3.6 By separating the themes out rather than keeping them combined, you can include more than one type of categorical data for a single grid cell. Note, for example, how the upper left-hand grid cell can now contain both roads and railroads.

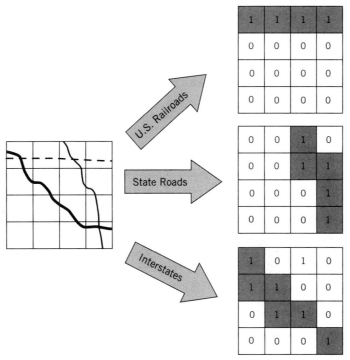

Figure 3.7   Coding ordinal data. Note how the different ordinal categories of road can also be represented by separate themes. Alternatively, they could be included as tabular data within a single theme by using the extended raster data model.

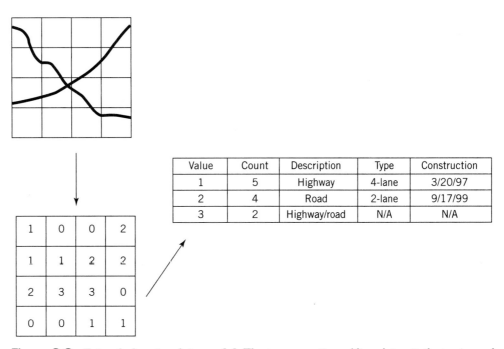

Value	Count	Description	Type	Construction
1	5	Highway	4-lane	3/20/97
2	4	Road	2-lane	9/17/99
3	2	Highway/road	N/A	N/A

Figure 3.8   Extended raster data model. The incorporation of line data attributes is made much easier by employing the tables within the extended raster data model. Here, we see an example of ordinally ranked linear data with additional descriptors.

line data. Under selected circumstances, any of the four basic systematic coding schemes can be used, as well as the unsystematic most important type method. Typically, polygonal thematic data do not overlap within a single theme, but this does not eliminate the problem of multiple themes occurring within a single grid cell.

CONCEPTUALIZING THREE-DIMENSIONAL SPACE WITH GRID CELLS

Statistical surfaces are often a fundamental portion of raster databases, especially, but not exclusively where terrain modeling is concerned. By definition, statistical surfaces do not include nominal or ordinal data. Most are ratio-level data, although temperatures in Fahrenheit or Centigrade scales are a classic example of interval scale surface data. These surfaces can also include such data as chronology, attraction between objects (e.g., gravity-model data), barometric data, and precipitation. In each case, statistical surfaces require that the data either be continuous or be at least assumed to be continuous, and therefore comprising an infinite number of points. This requires that the statistical surface must be sampled to be encoded into any computer tessellation, whether it is vector or raster. For raster encoding of surface data, the procedure is somewhat different than the five typical schemes we have already visited. Instead, one records a unique z value for each grid cell. The primary decision that must be made is exactly where, within the grid cell, the value is encoded. In most cases, either the centroid of the grid cell or one of the four corners is assumed. Although this decision will have implications for modeling, the important thing is to be consistent for all your data.

In most cases, surface data do not have attributes other than the z value for each sampled or interpolated grid cell value. Older, simple raster GIS software usually recorded a single integer value for each grid cell (Figure 3.9), whereas newer simple and extended raster models often record single rational values for each grid cell. Even in the case of extended raster models, additional attribute data are not normally included in additional tables. As we saw in Chapter 2 (Figure 2.5), the only values recorded are the X, Y, and Z coordinates.

Actually, statistical surfaces are often termed 2.5-dimensional rather than three-dimensional because they do not explicitly allow depth modeling. Scott (1997) has demonstrated that Map Algebra and raster data models can be extended to explicitly include volumetric data and information. Although these data models are currently somewhat experimental, it is likely they will become operational in the near future. For this text, we will abandon this extension of the statistical surface and the raster data model so we can focus on readily available data models and software. It is, however, important to acknowledge this innovation for future modeling capabilities, particularly with the advent of new object-oriented GISs.

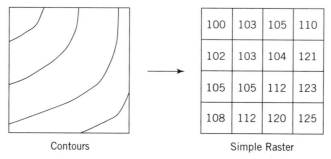

Contours   Simple Raster

Figure 3.9 Simple raster representation of elevation surfaces. Note the translation of contours on the left to discrete raster values on the right.

THINKING ABOUT THE MATHEMATICS OF MAPS

As we have observed, the most flexible and elegant conceptualization of the raster GIS coverage or theme is one of a series of numerical values arranged in rows and columns, as in the MAP model. More specifically, and largely on the basis of this model, we can view each raster GIS theme as a two-dimensional array of attributes, each represented by some mathematical value (or values, in the case of the extended MAP model), whose locations on the ground are implicitly encoded on the basis of the row and column position in the array. Moreover, each grid cell location for each additional theme must, to be of use in modeling, explicitly coincide with its column–row counterpart in the other themes.

A fundamental, intuitive understanding of this construct is absolutely essential for effective raster GIS modeling. All of the operators, functions, flow control procedures, and iterative techniques necessary to create and deliver models depend on it. This is the equivalent of understanding the chessboard, its red squares versus its black squares, and the rules imposed by that particular structure. Very soon, we will move to the operations and functions of the raster GIS on the basis of its imposed structure. To continue the chess analogy, this is equivalent to understanding the movements of the individual chess pieces and the rules, capabilities, and limitations of each piece. You would not begin playing chess without understanding both the game board and the pieces. Likewise, we will learn more about the GIS equivalents. And, just as with chess, we will eventually move forward to strategy, movement combinations, offenses, and defenses so that we will become first competent, then proficient, perhaps accomplished, and, with plenty of practice, eventually expert modelers.

A COMPARISON WITH AND CONTRAST TO MATRIX ALGEBRA

Map Algebra can be envisioned as the rules and operational procedures that are employed within the MAP raster data model and the capabilities it presents us. As we have already seen, map algebra is based on the fundamental MAP data model, especially on the two-deimensional array concept for each theme. In mathmetics, a two-dimensional array allows a set of mathematical procedures called matrix algebra to be applied to combine, compare, and manipulate the numbers of the matrix. Therefore, we can add matrices of numbers by taking each numerical value at each location for each matrix and adding it to its corresponding value. For example, consider the following matrix algebra equation:

$$\begin{bmatrix} 5 & 4 & 1 \\ 2 & 1 & 2 \\ 4 & 2 & 1 \end{bmatrix} + \begin{bmatrix} 3 & 2 & 1 \\ 1 & 4 & 5 \\ 2 & 7 & 3 \end{bmatrix} = \begin{bmatrix} 8 & 6 & 2 \\ 3 & 5 & 7 \\ 6 & 9 & 4 \end{bmatrix}$$

Notice how the upper left number (5) in the first matrix is added to the upper left number (3) in the second matrix to arrive at the upper left number (8) in the output matrix. The same can be said for the remaining eight numbers in each of the matrices. Now let's make a subtle change to the matrices to produce the following:

5	4	1		3	2	1		8	6	2
2	1	2	+	1	4	5	=	3	5	7
4	2	1		2	7	3		6	9	4

Notice how the matrices now have grid cells around them. This indicates that the matrices in matrix addition are virtually identical to the grid cells used in Map Algebra and that the process of adding matrices is identical to the process of addition

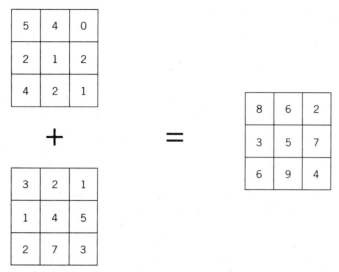

Figure 3.10 Map addition. Virtually identical to matrix addition, map addition adds each grid cell to each corresponding grid cell from two separate grid themes.

within Map Algebra. It could also be visualized with a more typical display of grid cell maps (Figure 3.10).

Just as matrix addition has its exact counterpart in Map Algebra, so does matrix subtraction. For each grid cell location in the first matrix, the corresponding grid cell in another matrix of numbers is subtracted to obtain the resulting values. And the same idea is again directly translated into Map Algebra, where each grid cell in one theme is subtracted from its corresponding grid cell in another theme.

If you have studied matrix algebra, you are aware that this one-to-one locational correspondence does not apply for such functions as multiplication, division, roots, and powers. This is where matrix algebra and Map Algebra part company. In Map Algebra, the locational one-for-one translation is maintained. Thus, to multiply the following simple 4 × 4 cell thematic maps within Map Algebra, we maintain the same rules that we applied to matrix addition and subtraction (Figure 3.11). The retention of this simple rule is essential because, unlike in mathematical matrices, the position of individual grid cells in raster themes directly corresponds to their position in geographic space. A result of this basic rule for our Map Algebra game board (our MAP-like data model) is that our grid cell values can be modified but their locations are not transposed or moved. All of the basic operators, functions, flow control, and iteration operations of Map Algebra and extensions to Map Algebra depend on this. Additionally, this knowledge is essential for those who use macro programming techniques to enhance and modify the basic data model and its user interface. We will discuss this topic later. For now, we will examine some of the basic operations available within software packages that employ some form of Map Algebra. As you go through the next sections, remember that each raster GIS approaches the use of Map Algebra differently. Try to keep the concepts rather than the commands uppermost in your mind as you learn how Map Algebra works.

AN INTRODUCTION TO MANIPULATIONS WITH MAP ALGEBRA

Despite its simple structure, or perhaps even because of it, Map Algebra is a very robust modeling language. Some form of it is employed in many well-known raster

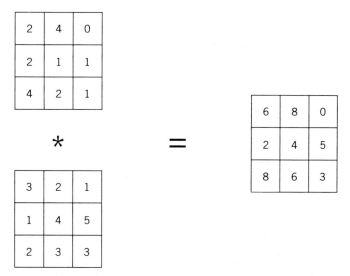

Figure 3.11   Map multiplication. Unlike matrix multiplication, in map multiplication the location of each grid cell is explicitly defined and the multiplication of two maps proceeds by multiplying each grid cell by its spatially coregistered grid cell.

GIS software packages, such as GRASS, ERDAS GIS MOD, ArcView Spatial Analyst, and ArcInfo GRID. Some will allow direct stacking of themes so that comparative Map Algebra calculations can be performed directly from theme to theme, whereas others simulate this process with the use of some form of Map Algebra calculator. The result is, for the most part, the same.

In our previous section, you might have gotten the idea that Map Algebra is simply a modified version of matrix algebra. Although this, in itself, is actually a major accomplishment in raster map manipulation, it is far from enough. Map Algebra is actually a complete modeling language, rapidly becoming the standard in the industry, that allows for program control, macro development, and iteration programming, as well as allowing for mathematical manipulation of theme grid cell values. In fact, even the analysis and manipulation of these grid cell values is not limited to mathematics but also includes a vast array of logical expressions that can be used to compare thematic values within single themes and among multiple themes. Thus, by combining elementary mathematical and logical operators into more complex functions and by using control and iteration, we can create complex models based on strategies that suit our particular data and modeling needs. Before we begin such complex modeling, we will first look at what types of operations are available to us.

Operators

The most basic functional characteristics of GIS packages based on the Map Algebra modeling language are the same operations with which we operate in most other modeling domains. As I have already suggested, this group of characteristics, called operators, can be divided into several groups—arithmetic, relational, bitwise, Boolean, combinatorial, logical, accumulative, and assignment. As you might guess, these include the basic functions often associated with formula translation computer languages such as FORTRAN (formula translation). Table 3.3 provides the typical sets of operators available.

TABLE 3.3   Operator Groups

Operator Group	Operator	Operator	Operator	Operator	Operator	Operator
Arithmetic	+	−	*	/	mod	...
Relational	<	>	==	>=		...
Bitwise	<<	>>				...
Boolean	&&	|	!			...
Combinatorial	and	or				...
Logical	in	diff				...
Accumulative	+=	*=	−=			...
Assignment	=					...

Arithmetic Operators   As you can see, there are many options for manipulating the grid cell values. The arithmetic functions include addition, subtraction, multiplication, division, and modulus (integers only), all required for constructing mathematical models within the raster GIS. Most of these operators will work equally well on integers as on rational numbers. The results of the operations will be based on the types of data used in the manipulations. For example, if only integers are used, the results will be integers. If floating point numbers are used in any of the operations—say, for example, floating point values multiplied by either integers or floating point values—the resultant numbers will be floating point. The only arithmetic operator that is limited by data type is the modulus (MOD) operator, which always returns integers. If MOD is applied to rational numbers, any remainder will be truncated and the result will be converted to integer. No matter which arithmetic operator you use, if there are cells without values (this assumes your GIS allows missing values or no-data values), the result will always be no data. Keep in mind that for most GISs, a zero contained within a grid cell is not necessarily the same as no data or missing data. Most often, the no-data category is contained within tabular cells in an extended raster data model. Figure 3.12 illustrates a typical use of the arithmetic operator based on multiplying two themes together.

Relational Operators   Relational operators are the kinds of operators you would normally expect to find within an RDBMS for use with its tables. In short, these operators evaluate a condition. If the condition is false, the output is assigned a 0, and if the condition is true, the output normally returns a 1. As with arithmetic operators, the no-data condition, when evaluated, results in no data. Conditions to be evaluated include *greater than, less than, greater than or equal to,* and *many more.* Relational operators operate on both integer and floating point numerical values and require at least two input values for comparison. Figure 3.13 demonstrates the use of the relational operator >= where it compares the numerical values of an input matrix to determine which of its grid cell values are greater than or equal to the values of the second matrix. Note how the no-data cells always return no-data cells after evaluation.

Boolean Operators   Boolean operators employ Boolean logic (true/false) and evaluate conditions, as we saw with relational operators. And as with relational operators, they return 1 for true and 0 for false; operate on grids, scalars, numerals, or combinations; and require at least two input values. No-data values are evaluated as no-data. Figure 3.14 shows a typical relational operator using the & (and) function, also known as the intersection. The figure demonstrates how grid cells that have val-

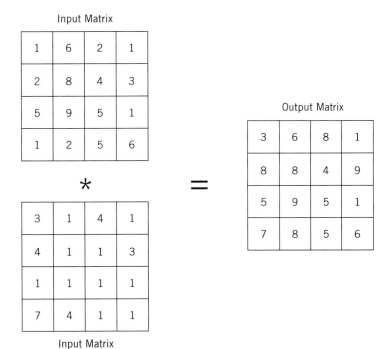

Figure 3.12   A 4 × 4 matrix example of map multiplication shows how two input matrices are multiplied to obtain the final result. This is functionally identical to the example in Figure 3.11 but shows that each set of values represents a map as a matrix of values.

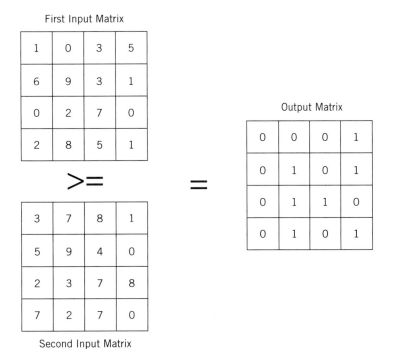

Figure 3.13   Relational operator. The grid cell values of the first input matrix are compared to those of the second. When the values of the first are larger than those of the second, a value of 1 is recorded in the output matrix. A value of 0 is recorded when this is not the case.

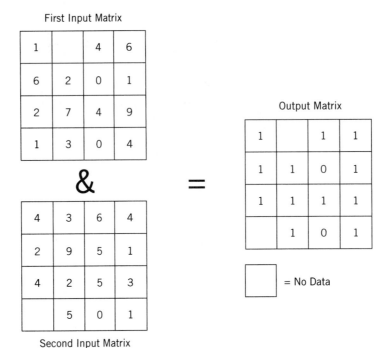

First Input Matrix

Output Matrix

& =

Second Input Matrix

= No Data

Figure 3.14 Boolean operators. The two input matrix values are compared to evaluate presence or absence of values in both sets of grid cells. When nonzero values exist in both, a truth value of 1 is recorded. If one or more corresponding grid cells contains a 0, a value of 0 is returned. Finally, when one or more values are missing (i.e., no data are present), the software returns no data.

ues in both themes are output as 1's whereas those that lack values in at least one of the two themes are assigned a value of 0. Note also the no-data output cells.

Bitwise Operators Bitwise operators compute on a binary representation of a single (input) set of matrix values and work only on integer values. If rational numbers are used as the input, they will be truncated first before they are evaluated, which means that the output values are always going to be integers. As with Boolean and relational operators, the results of using no-data values will always result in no-data output. Figure 3.15 demonstrate the use of the bitwise operator <<1, meaning all nonzero values will be converted to their binary equivalents. As you might imagine,

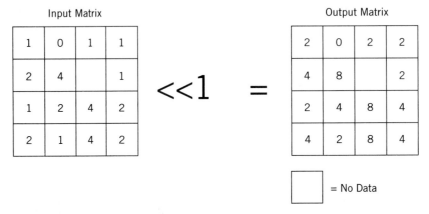

Input Matrix

Output Matrix

<<1 =

= No Data

Figure 3.15 Bitwise operators. Bitwise comparison of grid cells to a value of 1. Values of 1 are converted to their bitwise equivalent (2), values of 2 are converted to 4, and so on.

numbers in the input matrix that are equal to 1 will be converted to 2 in the output, 2 will be converted to 4, 4 to 8, and so on.

Combinatorial Operators Combinatorial operators share much in common with Boolean operators, except that they assign specific values to the results of evaluating two or more themes or matrices. This is a generalization of the Boolean operator, where a true condition is evaluated as a value of 1 and a false condition is evaluated as 0. In the case of the combinatorial operator, if both (or all) input values are evaluated as true (nonzero), the cell locations on the output grid are assigned some numerical value that preserves the uniqueness of the combination. So, for example, a 1 compared with a 2 would be assigned a different output value than a 1 compared with a 3. These numbers are asymmetrical also, in that a 1 on the first matrix compared with a 3 on the second will not return the same output value as a 3 on the first matrix compared with a 1 on the second matrix. Some older software (e.g., OSU-MAP-for-the-PC) allowed the user to specify the output values as the comparisons were made. To ensure that the unique results from analysis of asymmetrical input and output is preserved, in some software, such as ArcGrid, the assigned values are set on the basis of the order in which they are evaluated. Thus, for example, if the comparison of a 1 in matrix 1 with a 4 in matrix 2 is encountered before any comparison of a 4 in matrix 1 with a 1 in matrix 2, the 1 versus 4 is assigned a lower value than is a 4 versus 1. Table 3.4 provides an example of how this might be achieved. Notice that when similarly ordered value sets (e.g., 1 compared with 4) in one pair of grid cells is encountered again, the same value will be assigned. In this way, all identically ordered grid cell combinations will receive the same output value. Figure 3.16 shows a simple example of the application of combinatorial operators.

Logical Operators In addition to the Boolean operators we have already seen, there are some additional operators employing set-based logic. Three basic logical operators are generally included in this set—difference (DIFF), contained in (IN), and OVER. Each of these operators typically compares values based on pairs of matrices. DIFF compares two input matrices to determine whether the values in each matrix are the same or different. Although this is not universal, most software retains the first input matrix value if it is different from the matrix to which it is compared. If the

TABLE 3.4 Combinatorial Operators

Matrix 1	Matrix 2	Output
1	1	0
2	1	1
1	2	2
1	3	3
2	3	4
3	2	5
2	2	6
3	3	7

With combinatorial operators, the geographic information system modeler can decide what values are applied to each combination of grid cells. Each paired comparison is asymmetrical in that a 1 compared with a 2 is not the same as a 2 compared with a 1.

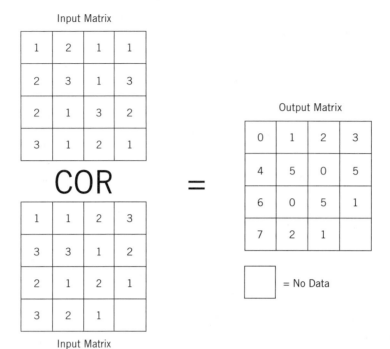

Figure 3.16 **Combinatorial operators.** The corresponding values of each grid cell are assigned unique values to represent the uniqueness of their combinations.

two matrix grid cell values are the same, the software returns a 0. In other words, in the output, all nonzero numbers indicate some change, perhaps a change from one time to another.

The IN operator, like the DIFF operator, accepts and compares two inputs, but they need not be grid maps. In most cases, the first input is an expression (typically a list or grid) and the second input is a set of numbers. The idea is to preselect a set of numbers against which you wish to compare the values in your grid matrix. If, for example you want to isolate several (let us say five) land uses coded with five individual integer values, and you further wish to zero out all others (sometimes called a mask in remote sensing), this technique is very useful. The output retains all values from the first input that are also contained in the second input (the set). Those that are not found in the second input are set to zero on output.

The OVER operator, which also accepts two inputs, searches for zeros. All nonzero values from the first input matrix are returned as output. If a zero is discovered, the software will return the second input value as output. This operation is similar to the IN operation, except that both inputs are matrices. Figure 3.17 is an example of the OVER operation.

Accumulative Operators The accumulative operators are designed for latter movement across a raster map, especially for cumulative surface analysis through scanning types of operations. A single grid is used as input and assigned an accumulative result to a scalar. If, for example, you were to use a += operator, the GIS starts at a corner (upper-left cell in a matrix is used by GRID), then moves to the first cell to the right and adds the value of that cell to the value of the first cell. Then the scan proceeds to the next cell and adds its value to the sum of the values of the first two cells. The process then continues until you run out of grid cells. As with other operators we have seen, no-data cells are ignored; in this case, they are not used at all for accumulation calculations. Figure 3.18 illustrates the += accumulative operator case.

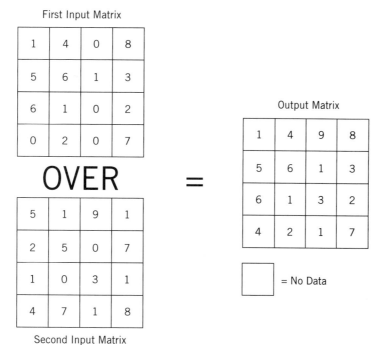

Figure 3.17 **Logical operators.** The OVER logical operator returns the value from the first input matrix unless a 0 is encountered in the first matrix, in which case the second grid cell value is returned.

Assignment Operators The final group of operators is the assignment operators. These store the results of expressions in an output (normally a grid cell matrix). This operator is actually as simple as assigning all cells of an input grid to a single value. More complex mathematical expressions and arithmetic operators can also be included. So, for example, an output matrix could be created by multiplying one theme by another, or an output grid matrix could be created by multiplying a grid matrix by a single value (e.g., an input grid matrix multiplied by 5). This latter case is demonstrated by Figure 3.19. Additionally, tabular values from an extended raster model could also be used to manipulate the raster values in a separate grid.

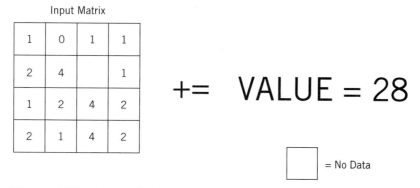

Figure 3.18 **Accumulative operator.** The values of all grid cells are added to arrive at a single numerical value. In this case, all grid cell values added equal 28.

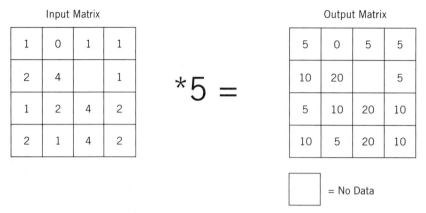

Figure 3.19 **Assignment operator.** These operators allow the user to assign values based on their own requested output. In this case, the operator multiplies each grid cell value by 5. The output is equivalent to creating a second output grid filled with 5's.

Functions

Functions are higher-order GIS operations built up of the more basic operators we have just examined and designed to provide a vehicle for model implementation. Functions are grouped as local, focal, block, zonal, global, and special types. In the next chapter, we will look more closely at the functions; however, we will provide some basic concepts and definitions here as important building blocks of the Map Algebra modeling language.

Local functions, also called by-cell functions, are designed to operate on a cell-by-cell basis. In other words, each grid cell at a given location in a first matrix is operated on either by an expression or by a grid cell at a corresponding location in another matrix.

Focal functions, or by-neighborhood functions, characterize or reclassify a selected cell on the basis of characteristics of a predefined neighborhood of grid cells.

Block functions, or by-area functions, are similar to focal functions in that they do evaluate groups of grid cells to perform its reassignments. In this case, however, the results are assigned to whole blocks of cells, and the groups of cells do not overlap.

Zonal, or by-zone, functions use zones identified from another coverage to evaluate and reclassify a target cell. The zones are actually geographic areas, whether contiguous, fragmented, or perforated, and are usually defined by internal attribute homogeneity.

Global functions, as the name implies and unlike the previous types of functions, tend to operate on the entire grid matrix all at once. These are the types of functions most often associated with Euclidean geometric analysis, distance and shortest path analysis, and visibility and viewshed analysis.

Beyond these are an even more complex set of special functions, often based on selected integration of simpler functions. These specialized functions are primarily used for such functions as complex geometric analysis, hydrological modeling and characterization, and surface analysis and characterization.

Not all GIS software contains all of these types of functions, but most professional raster GIS software does contain macro language capabilities for implementing these capabilities. As you might already suspect, the existence of a MAP-like data model is useful, if not essential, for optimal model implementation. Before such implementation can take place, however, it is necessary to be able to issue commands that control the flow of operations. This flow control functionality is explained next.

Flow Control

Flow control is an integral component of Map Algebra. It provides for basic functions such as starting and stopping, and for a set of conventions that can be combined to create a command line framework within which the user can interface with the GIS software. Although an early variant of this framework was first developed by Tomlin and Berry (1979) for the original Map Analysis Package, it has been modified to a lesser or greater degree for all of the variants of this original data model. As such, the exact nature and structure of the flow control framework will vary from package to package. We will adopt the same format employed by Tomlin (1990).

Whether your GIS software uses a graphical user interface (GUI) or is strictly command-line driven, the process is essentially the same because the commands and structure of the language are embedded in the GUI. The basic format for flow control is composed of two distinct, yet linked elements—statements and programs—that work with the operations and functions we have already seen. These two elements are linked in a hybrid language that resembles elements and structure of both algebra and English. Some non-English versions of this flow control language also exist.

Statements A statement is a verbal representation of the operations. It provides a declarative command structure that links operators, functions, and programming commands in a logical progression. Much like declarative computer languages, the order of operations is vital to proper functioning of the model. In fact, some GIS packages use a set of flow control procedures that resemble such computer languages as FORTRAN or versions of C or BASIC. For ease of use, I will describe the typical form of the language that more closely resembles a declarative English sentence. Because of its flexible, natural language–like structure, this flow control methodology can also be extended to create algorithms in higher-level GIS macro languages.

The statements used in the flow control methodology include an ordered sequence of letters, numbers, symbols (representing operators), and blank spaces. As with a declarative natural-language sentence, the sequence forms a declarative statement that indicates the subjects under consideration, modifiers to the subjects, and objects on which the subjects will act. Consider the following statement:

TotalCostMap = LocalSum of FirstCost and SecondCost and ThirdCost

This statement begins with the subject. (For most modern GISs, this would be the name of output maps.) In this case, the subject is TotalCostMap. This is the thematic map or other output that will result when the operations on the right of the equals sign are executed.

Statements also have the ability to accept modifiers that correspond to prepositions, adjectives, adverbs, nouns, conjunctions, and even punctuation that add significance or change the meaning of the sentence. The equal sign in the statement above is a modifier because it acts as a verb providing action on the subject. Other modifiers include the preposition *of* and the conjunction *and.* These words provide meaning, enhance meaning, and link other terms in the statement just as they would in a normal sentence.

The objects *LocalSum, FirstCost, SecondCost,* and *ThirdCost* provide items and actions on those items that, together, create the output (the subject). In this case, the *LocalSum* term is a function (a local function, in fact), and it operates on each cost grid cell for each of three separate matrices by adding them together. For most Map Algebra–based raster software, the objects may be map names (as in the case above), nouns, adverbs, or numbers. Special codes are also possible that represent special values such as the null set (–0), the highest value (++), the lowest value (– –). The code

... can be used to indicate a missing portion of a numerical series, such as *1 2 ... 8 9 10*, which indicates that the sequence values *3 4 5 6 7* are part of the arithmetic series.

In some cases, the actual values of portions of a Map Algebra statement are either not known or will be evaluated, and therefore created, during evaluation. These portions are called variables and are analogous to variables in a traditional programming language. In the traditional implementation of Map Algebra, these variables are noted by all-capital letters. Thus, a statement could look like this:

NewMap = LocalSum of FIRSTMAP and SECONDMAP and THIRDMAP and FOURTHMAP

Where the objects *FIRSTMAP*, *SECONDMAP*, *THIRDMAP*, and *FOURTHMAP* suggest the nature of the values that will, on evaluation, replace the generic variables. This generalized statement allows for flexible algorithm development and implementation no matter what the actual values will be.

To add more flexibility to the programming environment, Map Algebra also allows the inclusion of optional portions of a statement. This might be required if, as in the previous statement, we know that at least two of the maps will be needed in the LocalSum operation whereas the remaining maps are optional and need be evaluated only if they are present. The modified statement often uses brackets to define which of the portions are optional. So the previous statement would be rewritten to look like the following:

NewMap = LocalSum of FIRSTMAP and SECONDMAP [and THIRDMAP and FOURTHMAP]

The assumption from the statement above, however, is that there are four potential maps to be evaluated and that only the first two are required. What if you are not aware of exactly how many maps will be included in the LocalSum operation? Fortunately, Map Algebra allows for this with still more flexibility by adding the term *etc.*, which indicates that the statement portions enclosed within the brackets can be continued. Rather than explicitly including the variables *THIRDMAP* and *FOURTHMAP* in brackets, we use a more generic variable such as *NEXTMAP*. The statement for NewMap based on the LocalSum can now be rewritten thusly:

NewMap = LocalSum of FIRSTMAP and SECONDMAP [and NEXTMAP] etc.

This shows that any number of maps beyond the first two can be included in the calculations in addition to the mandatory first two.

As you have observed from the statements above, we have been using the term *LocalSum*. From our preliminary discussion of functions, you should recognize that this is a member of the group of functions we called local or by-cell. Any function, no matter which group it belongs to, can be incorporated into our statements to create procedures that will evaluate our grid maps. In the next chapter, we will look more closely at the functional options available in Map Algebra–based GIS packages. Before that, however, we need to briefly examine programs and iteration.

Programs The notational representation of a procedure in Map Algebra is called a program. Although this sounds like the same thing as a statement, it is not. Rather, a program is an ordered sequence of statements, each statement placed on a separate line and collectively designed to perform a wide array of interconnected activities. You should note that I used the term *ordered sequence*. This indicates that, just as in any other programming language, the order in which the statements appear will most often indicate the order in which the processing takes place. The program is the notational specification of a GIS model and indicates which maps will be included, what operators will be applied to each specified grid cell or cells, what intermediate maps will be produced, and how they in turn will be manipulated. You should also note that I stated that the order of statements will *most often* indicate the processing order. This would assume that the program is strictly linear and that each step directly stems from the one previously implemented before it. This would limit our programs to very simple tasks. We need to add one more feature to our Map Algebra language to provide us with more flexibility for modeling in complex situations.

Iteration

Fortunately, Map Algebra is a robust language allowing the GIS modeler or programmer to vary the order in which statements occur. For example, there are steps that one might wish to skip under specified circumstances. Thus, if the evaluation of a given statement results in values that do not achieve a specific threshold value for our model (say, a threshold value for total erosion on a soil map in an evaluation of housing sites), we might be able to eliminate procedures that would otherwise be used to add financial resources to make development economically viable. We might also have a need to insert some operations under given situations. In fact, we could rewrite our Map Algebra program of housing site development in such a way that we would normally assume that the need for additional financial input would not be encountered, and if the erosion threshold is exceeded, we could then ask the program to include the necessary statements and procedures to implement the addition of financial resources. Essentially, this is the reverse of the first situation. These two situations are akin to an "if-then-else" type of programming statement available in nearly all modern programming languages.

There will also be common situations in which some procedures and statements will need to be repeated to achieve the final result. This is often the case when statistical or numerical procedures require multiple steps for their final evaluation. In the case of map algebra, a classic example would be a model in which many time steps are to be processed. Each successive step must be performed and its intermediate map output stored and then retrieved for the next time step. The computer language analog would be a "do loop" structure that requires that operations continue until some predefined stopping point or until the data set is exhausted.

Finally, you should think of Map Algebra as a complete high-level spatial modeling language that includes basic elements (operators), more complex elements (functions), and a formalized structure (statements), together with all the necessary programming features that allow complex models to be developed and implemented. To become proficient at GIS modeling in raster, it is essential that you become familiar with the structure, operating rules, and components of the individual version, implementation, or modification of Map Algebra that your particular software employs. We will look at some of the more powerful functions available within the Map Algebra language in the next chapter. Be sure to compare these to those available with your software before you begin modeling. A thorough knowledge of the modeling capabilities of your software will improve efficiency, inspire more ideas, and suggest new, as yet unimplemented capabilities that will take you beyond the GIS button bar into the exciting and potentially lucrative realm of applications development.

Chapter Review

Point, line, area, and statistical surface attribute data can be conceptualized and represented as nominal, ordinal, interval, or ratio scales of geographic data measurement. Each of these measurement scales provides both opportunities and restrictions as to how they are stored either within simple raster models or extended raster models. The dimensionality, the associated measurement levels, and the potential for locational overlap among objects all contribute to determining which of the five raster input methods (four systematic and one unsystematic) would be the most appropriate. Statistical surface data generally do not have additional attribute values associated with them and are also the most common data represented as floating point (rational) values rather than integers.

The raster tessellation amounts to a matrix not unlike what one might encounter with matrix algebra. Within this construct, most modern raster GIS packages have adopted a modeling language as some variant of Map Algebra. Map Algebra is a formalized, simplified version of matrix algebra that maintains the locational fidelity of each grid cell within the matrix. Beyond simple mathematical procedures, Map Algebra also includes a wide array of relational, logical, combinatorial, accumulative, and assignment procedures. Collectively, these are called operators, which can be combined with higher-level procedures called functions, within a natural language-like structure called statements, to allow program control to build programs used for GIS model development and deployment.

Map Algebra also allows for flow control through ordered sequences of Map Algebra statements called programs. The sequencing of program statements imposes a framework on which maps are selected and operated on in which order to achieve the desired model outcomes. Additional programming flexibility is incorporated through the use of statements that allow some procedures to be skipped, included, or iterated whenever necessary. This flexibility gives Map Algebra the same power and flexibility most often associated with typical computer programming languages.

Discussion Topics

1. What impact does dimensionality of objects have on the selection of appropriate raster encoding scheme?

2. What is nonsystematic raster encoding, and what types of criteria might you employ in selecting and implementing it?

3. Using a hypothetical or a real set of map themes, discuss which themes would appropriately use the extended raster data model and provide some concrete examples.

4. How does the raster encoding methodology selected impact the spatial accuracy of point, line, and area entities?

5. Discuss the impact of point, line, and area objects that occur within the area occupied by a single grid cell. Describe some situations in which this is likely to occur and explain some solutions to the problem.

6. How do the mathematics of Map Algebra and matrix algebra differ? Why are the mathematics of Map Algebra different? Couldn't we simply have added some of the Map Algebra structure to the mathematics of matrix algebra?

7. List and provide a simple, one-sentence description of each of the basic types of functions.

8. Describe some different types of operators and functions and provide examples of how they operate within the statement structure of Map Algebra.

9. Provide examples of statements that illustrate the different aspects of flow control. In your statements, include variables, objects, modifiers, and other statement

parts. When these are completed, label the parts of the statements, much as you would diagram a sentence for an English class.

Learning Activities

1. In this chapter, we learned five separate ways of encoding raster data. Create a coding system in which each code indicates one of these. For example, *PA* could stand for *presence/absence*, *DT* for *dominant type*, etc. Now create a table that shows the dimensions of the geographic entities on the vertical axis and the encoding scheme methods on the horizontal axis. In each cell, place an X where the coding scheme could be used for each dimension of data type.

2. Provide five examples of how the most important type methodology of grid cell encoding could be employed for raster data. Be specific about how you made your decisions—exactly what basis you are using to decide what the most important type is.

3. Create, or copy real-world examples from available GIS databases, of extended raster database management tables for data of the following types:

   a. Land use example (polygons)

   b. Linear infrastructure example (power lines, street networks, highways, etc.)

   c. Point examples (wildlife, stores, wells, etc.)

4. Pick up a copy of a linear algebra (matrix algebra) text and illustrate examples of matrix multiplication, division, square root, and square. Use real numbers to solve these problems. Now, create illustrations of Map Algebra multiplication, division, square root, and square. Again, use real numbers (identical to those for the matrix algebra examples) and work through the solutions. Describe the results to illustrate the differences between the two.

5. On the basis of the introductory material you have been given in this chapter, create simple examples of Map Algebra statements that include operators and functions to derive the following output maps. The titles of the output maps are meant to be descriptive of the methodology you need to employ. Remember, the output map names are only descriptive. Your results may vary depending on how you interpret what the output map means.

   a. BiggestMap =

   b. SmallestMap =

   c. AverageMap =

   d. DifferenceMap =

   e. TimeChangeMap =

6. Create a simple, fictitious Map Algebra program (an algorithm, because you are not using real data yet) that includes at least three of the statements you just created in question 5 above and any others you wish to use. Describe what the program is doing and what its output is meant to represent. For an extra challenge, try adding at least one flow control statement that allows iteration or condition and also includes the use of a variable.

## CHAPTER 4

Characterizing the Functional Operations

On completing this chapter and combining its contents with outside readings, research, and hands-on experiences, the student should be able to do the following:

1. Define and provide graphical examples of local GIS operations based on a variety of mathematical and logical operators and be able to evaluate these simple examples

2. Using available raster GIS software, implement the examples you solved manually as well

3. Explain what is meant by the term *worm's-eye view* when applied to local operations

4. Define *focal operations* and explain the difference between focal operations and local operations

5. Define, provide graphical examples, and manually solve focal GIS functions and employ your GIS software to solve these same examples

6. Define, provide graphical examples, and solve block, zonal, and focal functions both manually and with your raster GIS software

7. Implement specialty geometric, hydrological, multivariate, and surface evaluation functions both manually with sample data sets and with your raster GIS software

8. Begin placing operators and functions within Map Algebra statements to create complex algorithms and implement the algorithms using your raster GIS software

9. Examine implemented models (either your own or those of others) and dissect them into the parts from which they were created

FUNCTION REVIEW

As you may remember, functions are higher-order operations than their more elemental components that we have already defined as operators, but they are less complex than either statements or programs. We have seen that functions come in a variety of categories, each of which operates in a set of unique ways. Perhaps the most important individual concepts that are needed to become effective as a GIS modeler are the functions. Although statements and programs often affect the model control, the functions define what a raster GIS is capable of doing within the structures imposed by statements and programs. Every raster GIS software package is different and contains an equally unique number and character of functions. The following discussion will provide a framework for understanding and should give some idea of the general functional capabilities that you can expect to be available within a professional raster GIS package that implements some variant of Map Algebra.

Local Functions

As we have seen, local or by-cell functions operate at a very local level. By local level we mean that the focus is on the individual grid cell, whatever resolution you have chosen to encode it. In this way, if your grid cells are 100 meters on a side, the local function operates on a 100-by-100-meter space. If, on the other hand, you have grid cells that have a resolution of 10,000 meters on a side, the amount of local area operated on is 10,000 by 10,000 meters. As you might guess, the impact of calculations through local functions is nearly entirely dependent on this resolution. Whatever your resolution, the local function can be thought of as giving you what Tomlin has called a worm's-eye view. What this means is that if you were at ground level, such as a worm might be, your view of the world would be limited to a very short distance from your absolute location, or, more to our point here, your view of the raster GIS theme is limited to your immediate grid cell.

Most GIS models are not limited to a single layer or theme, however, so we have to extend our idea of the worm's-eye view to vertically coregistered grid cells. Thus, our worm is also able to look up at data directly above and directly below its specific location. Local functions then begin at local grid cell locations on a single theme and are manipulated on by either individual operators or coregistered grid cells from other themes (Figure 4.1). Notice from Figure 4.1 that you begin with each individual target cell and manipulate it through available operators to obtain an answer specific to that same cell location.

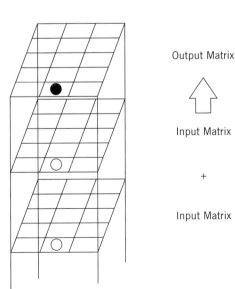

Output Matrix

Input Matrix

+

Input Matrix

Figure 4.1   Local function. Local functions are cell-by-cell functions that compare each individual grid cell from one matrix with its corresponding grid cell in the second and all succeeding matrices.

Despite their seeming simplicity, local functions are both potentially robust and very commonly applied to modeling tasks. In fact, they are among the most common functions applied in raster GIS modeling. Let's take a look at some of the available options for local functions analysis. Remember that all of the Map Algebra operators can be applied under given circumstances, so we will consider local functions by the operators they might use and illustrating the results.

The potential operators most commonly applied to local functions can be grouped into six categories:

1. Trigonometric

   • Trigonometric: sin, cos...

2. Exponential and logarithmic

   • Exponential: exp, exp10...

   • Logarithmic: log, log10...

   • Power: sqrt, pow...

3. Reclassification

   • Reclassification: reclass (renumber)...

4. Selection

   • Selection: select, selectcircle...

   • Condition: con, test, pick...

5. Statistical

   • Statistical: min, mean, majority...

6. Other

   • Arithmetic: abs, ceil, rand...

It might be difficult initially to imagine what each of the dozens of local functions might be used for. You might be asking yourself, "What possible use could I make of power functions, cosines, logarithms, and so on, for modeling?" Individually, these functions might not be particularly useful, but if you remember that a model composed of a single local function is not only unlikely but seldom very useful, you might anticipate that these functions are typically only components of larger models. These models might take the form of, for example, a regression equation. In such a situation, the power function is going to be employed as each grid cell's contents becomes part of that equation. This will allow you to perform the regression on a series of themes at one time. Some of the themes will be composed of constants, some of calculated values, and still others of variables whose values will change with other inputs. The following discussion will show you the structure and methodology of these functions. It will not provide all possible situations in which each function will be integral to a larger model. That would require a nearly infinite number of possibilities. I leave it up to you to envision how these and the other functions might be used. Later in the book, I will provide some model examples that might assist you in your own work.

So having classified local functions into the six groups and their smaller categories, we begin our examination with trigonometric local functions. These functions perform trigonometric calculations on cells of raster maps or on numbers or scalars. When applied to raster maps, they can operate on a single grid cell, on groups of cells, or on all grid cells of the map (Figure 4.2). As you might guess, the output from trigonometric functions would be floating point values. For this reason, raster GIS

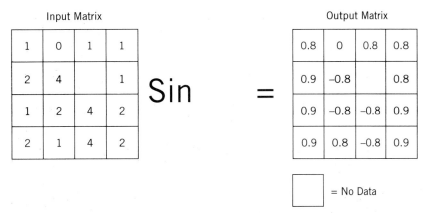

Figure 4.2 Trigonometric local function. Here, a sine function is applied to each grid cell.

software that does not support rational mathematics either will truncate the output values or will not implement these trigonometric functions.

The exponential and logarithmic functions are nearly identical in operation to the trigonometric functions, except that they perform log and exponential calculations and the power function requires an input describing what power value (e.g., squared, cubed). These functions should include all the powers, roots, and logarithms (including natural logs).

Perhaps the most used group of functions is the reclassification group of local functions. Offering a wide range of options because they are totally under the user's control, reclassification operations allow users to select either individual cells or groups of cells for consideration to be recategorized. We will examine how the selection function can be used in conjunction with the reclassification function to do this. If you have worked with vector GIS software, you might remember using some form of reclassification and line dissolve to combine nominal subcategories such as *housing*, *business*, and *industry* into a larger category like *urban/built-up*. You might also wish to reclassify nominal land cover or soils categories on the basis of their ability or relative availability to support selected land uses (Figure 4.3). In this way, you are converting the grid cell values from their original nominal data measurement level to rank order (ordinal). This should also suggest the ability to use the reclassification function to weight both individual grid cells and whole grids. This same process can take place in the raster environment by reassigning grid cells in a simple raster data model or creating new values in a remap table of an extended raster GIS model.

The second example use of reclassification above shows how it allows the capability of changing the geographic data measurement scale of some or all grids (themes) in a database. On the surface, this may seem fairly innocuous, but it can have extreme consequences on the validity of models that are derived from its application. These consequences can be either positive or negative, depending on how the measurement scales are changed and how the recalibrated grids or values are used. If, for example, you were to convert soil type data (nominal or categorical data) into land capability weights (interval or ratio) so that they could be compared mathematically with interval or ratio data, indicating a reclassification of existing land use types into land suitability weights, the results would be of no use. A frequent, and fatal, mistake is to use local functions to compare nominal with ordinal, interval, or ratio grid cell data. For example, it is possible with raster GIS to multiply land cover grid cell numbers (numbers representing nominal categories) by, for example, elevation values (on the ratio scale). The result may often be visually compelling, but the values are meaningless. In short, the frequency of use and the power of local functions should suggest that extreme caution be used when applying them.

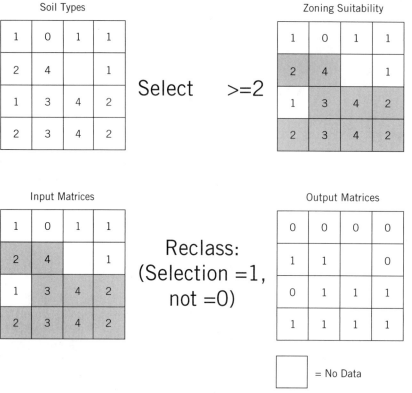

Figure 4.3 Reclassification local function. This example shows how soil type values are compared with zoning suitability values from a second matrix. When both pairs of values from each corresponding matrix are greater than or equal to 2, a value of 1 is returned. A value of 0 is returned if this is not the case; a no-data class is returned when data are not available for comparison.

To be totally effective, reclassification functions need to have the capability to isolate more than one grid cell, and often a subset of the entire grid. The selection function allows for the identification, isolation, and subsequent manipulation with other functions, what is often a subset of an entire grid. The selection can be performed on the attributes of the grid cells where, for example, you could isolate all values that are the same (e.g., all values = 6), or all values could be within a specified limit (e.g., all values between 3 and 6 [Figure 4.4]), or all values could share the same descriptor

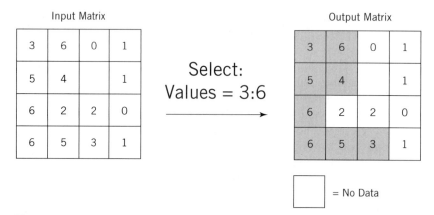

Figure 4.4 Subset reclassification in a local function. This example shows how all values between 3 and 6 are selected for reclassification (shaded), whereas the remainder are not evaluated.

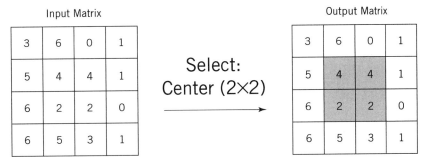

Input Matrix

3	6	0	1
5	4	4	1
6	2	2	0
6	5	3	1

Select:
Center (2×2)

Output Matrix

3	6	0	1
5	4	4	1
6	2	2	0
6	5	3	1

Figure 4.5   Positional local reclassification function. Another method of deciding which grid cells to reclassify is by defining the position of the grid cells to be examined. In this case, the software is directed to go to the center of the matrix and select a 2 × 2 matrix of cells. Many other methods of selection are also available.

(e.g., all grid cells are classified as corn). These attribute-based searches are performed either on the values of the grid cells themselves if a simple raster data model is used or on the tabular data in an extended raster data model. Depending on your software, the selection of grid cells by attribute may require human intervention to make the selections or may include some form of test function that automatically searches the grid cells or attribute tables and compares what it finds against a set of criteria. The more automated the system, the more easily modeling can take place.

Another approach to selection is to select grid cells on the basis of their individual or collective positional information. You could, for example, select all corner grid cells, the central grid cell, the leftmost column of grid cells, or the top row of grid cells. These are pretty simple examples. Other types of selection functions could include the isolation of other geometric shapes, such as boxes (Figure 4.5) and circles. Selecting such geometric shapes requires that selection functions also allow the inclusion of techniques for defining exactly where, in geometric space, the grid cells are located. Graphics software searches provide some additional examples of how features (in this case, vector features) can be selected. You may be familiar with such terminology as *connected to*, *within*, *inside*, *outside*, and *like*. These same types of statements can also be used to select sets of grid cells from a grid. There should also be some method of linking both attribute-based and locational-based selections.

Statistical local functions are primarily designed for comparing two or more input grids. You may return, for example, the minimum, maximum, mean, median, majority, or minority value for each cell location. Beyond comparing one grid with another, you can also make comparisons with constants or other numerical values. The output of the statistical local functions is usually a grid (Figure 4.6).

Under the final group of local functions, "other," the primary type of functions are arithmetic functions, based largely on some form of arithmetic operators. Among the most common local functions in this group are those that allow the reassignment of floating point grid cell values to integers and vice versa, those that find absolute values of numerical grid cell values, and those that allow the assignment of random numbers to grid cells. We have already seen the importance of the ability to manipulate geographic data measurement so that the levels match from one grid to another. The same is often true of integer versus floating point mathematics, especially if large databases are used in modeling where floating point mathematics might slow the model down too much for it to be effective. Additionally, the ability to change negative numbers to absolute values is quite useful if your grids are designed to indicate some attribute magnitude but not necessarily a direction. Random number generation is vital to some modeling processes, especially those that are attempting to predict future events, as might be employed through a Monte Carlo simulation model of urban growth (Meaille and Wald 1990), the spread of fire (Liu

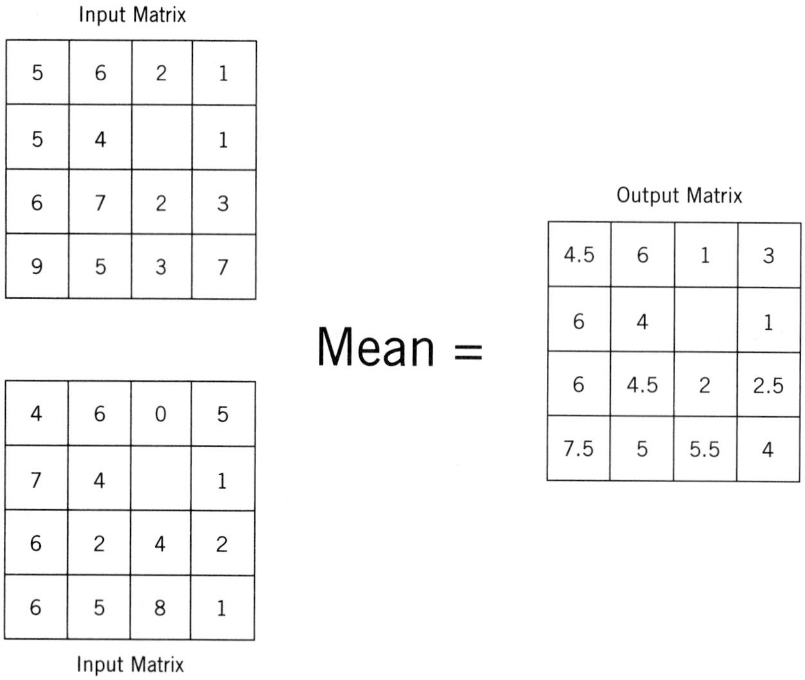

Input Matrix

5	6	2	1
5	4		1
6	7	2	3
9	5	3	7

Mean =

Output Matrix

4.5	6	1	3
6	4		1
6	4.5	2	2.5
7.5	5	5.5	4

4	6	0	5
7	4		1
6	2	4	2
6	5	8	1

Input Matrix

Figure 4.6 Statistical local function. Many statistical techniques allow for comparison of groups of coregistered grid cells. Here, an arithmetic mean is applied to two input matrices.

1998), or other diffusion processes (Mattikalli 1995, Miyamoto and Sasaki 1997, Park 1996, Portugali et al. 1994).

Perhaps among the most important local function in this group is one that allows for evaluating conditions in a grid. In ESRI's GRID module, this conditional statement (CON) is explicitly included in the function set. The CON can link multiple grids and compares the conditions of these grids on a cell-by-cell basis. The CON command can be applied to multiple conditions at once, but each expression or value must be able to assign values to the grid cells. Results of evaluation of the conditions will normally result in a true expression, which is a predefined numerical value for the conditions being met (Figure 4.7). A false expression can also be assigned within the function so that specific output values can be assigned to that condition should it be encountered. In the case of GRID, if no value is assigned to the false condition, this results in the assignment of no-data to those grid cells. One thing to keep in mind with the CON command is that it permits the creation of only a single output grid. GRID includes an IF statement in its Map Algebra language that allows for the construction of conditional evaluations that can produce multiple output grids. Both are very useful if complex models are to be developed that require the evaluation of conditions. We will revisit both the conditional function and the IF statement later on in the text.

Focal Functions

Unlike local functions, focal functions move beyond the worm's-eye view to evaluate the grid cell values in our raster map themes, although they do share the evaluation of individual grid cells with the local functions. Focal functions compute the output map or grid by assigning the values of output grid cells on the basis of some function of the input cells of some specified neighborhood of input grid cells of the source

First Input Matrix 1

4	6	2	5
7	4	7	1
2	5	4	2
6	5	8	1

Con:
Matrix 1 >= Matrix 2
(True = 1, False = 0)

Output Matrix

1	1	1	0
0	1	0	1
0	1	0	1
1	1	0	1

4	6	0	7
6	4	9	1
6	2	5	1
3	5	9	1

Second Input Matrix 2

Figure 4.7 Conditional local function. A comparison of two matrices allows the user to return true (1) or false (0) responses as indicators of the condition being compared. In this case, the condition being examined is whether the first input matrix is greater than or equal to the second.

location or grid cell. In other words, we now start at a single grid cell, look at its neighboring grid cells, and analyze the contents of those neighbors to derive the value we will assign to our output grid cell (Figure 4.8).

Focal or neighborhood functions provide a wide range of possible neighborhoods from which to begin an analysis. Some typical forms are rectangles, circles, annuluses (doughnut shapes), and wedges. These can also be selected by size. For example, one could have a 3 × 3–cell rectangular neighborhood, a 6 × 3–cell rectangle neighborhood, a circle of radius 6 cells, a wedge of 9 cells, and so on. The placement of these shapes is largely dependent on target cell within the selected neighborhood.

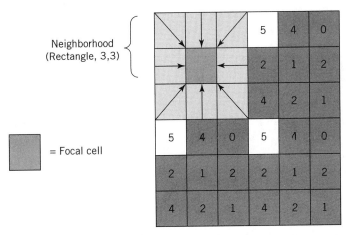

Neighborhood
(Rectangle, 3,3)

= Focal cell

Figure 4.8 Focal function. In focal functions, we examine the target cell and its neighbors and return a value based on their evaluation.

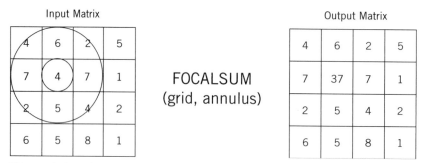

Figure 4.9 Annulus neighborhood. In an annulus (a doughnut shape), we evaluate the cells contained in the ring but not the target cell itself.

In the case of an annulus, the target cell is placed at the center of the doughnut shape, and the results are not based on the attributes of the target cell at all, because it is not part of the doughnut neighborhood (Figure 4.9).

Professionals in remote sensing are familiar with high- and low-pass filters. These are essentially moving windows (neighborhoods of cells). Moving filters are another type of focal function, in that a window is created and its output values are assigned, one at a time, until the entire output grid is filled. As with moving filters found in remote sensing packages, moving focal functions allow the introduction of kernal values or weights within the neighborhood. Usually the weights are stored as a separate file to be included in the local function statement. As with moving filters, the weights can be uniform, symmetrical, or asymmetrical. You might, for example, wish to assign greater weights to grid cells near the target cell and lower weights as you move away. In this way, you could simulate a distance weighting function for the grid cells, resulting in a distance weighting effect in your output.

Beyond the ability to change the size, shape, and weights of focal functions, there are many options for manipulating the contents of the selected neighborhoods for evaluating the target cells. In fact, most software using the Map Algebra syntax will allow you to modify the method of processing the neighborhood cells to suit your own needs. The general forms of focal function include such methods as sum, majority (e.g., focalmajority), minimum, maximum, mean, median, range, standard deviation, variety (or diversity), and flow. Let's examine just a few of these so you begin to understand how they work.

A focal majority function evaluates all the grid cells in an input grid neighborhood to determine the majority of the values enclosed within the specified neighborhood and returns that value to a coregistered target cell in the output matrix. So if you selected a 3×3 rectangular neighborhood focused on the central cell in that matrix, and the majority of the grid cell values were 2's, the result would be to place the number 2 in the target cell of the output grid (Figure 4.10a). In the same grid, we could perform a focal minimum that would result in the value of 1 being passed to the output grid (Figure 4.10b). Or we could average the nine values (i.e., focalmean) to return the average value to the output grid cell location (Figure 4.10c). A final example would be to evaluate the diversity or variety of categories in the neighborhood, in which case the function would determine how many different types of grid cell values there are and would return that numerical value to the target cell in the output grid. As you can see, there are many ways in which these functions can be employed to characterize the grid cell values in each neighborhood. These could be employed to determine, for example, the minimum cost of houses in a neighborhood, the total worth of land in the neighborhood, the average number of crimes in a neighborhood in a given year, or the landscape level diversity of land cover types in a patch (neighborhood).

One additional and quite useful application of focal functions has to do with the flow or dispersion across or through neighborhoods. You might envision the move-

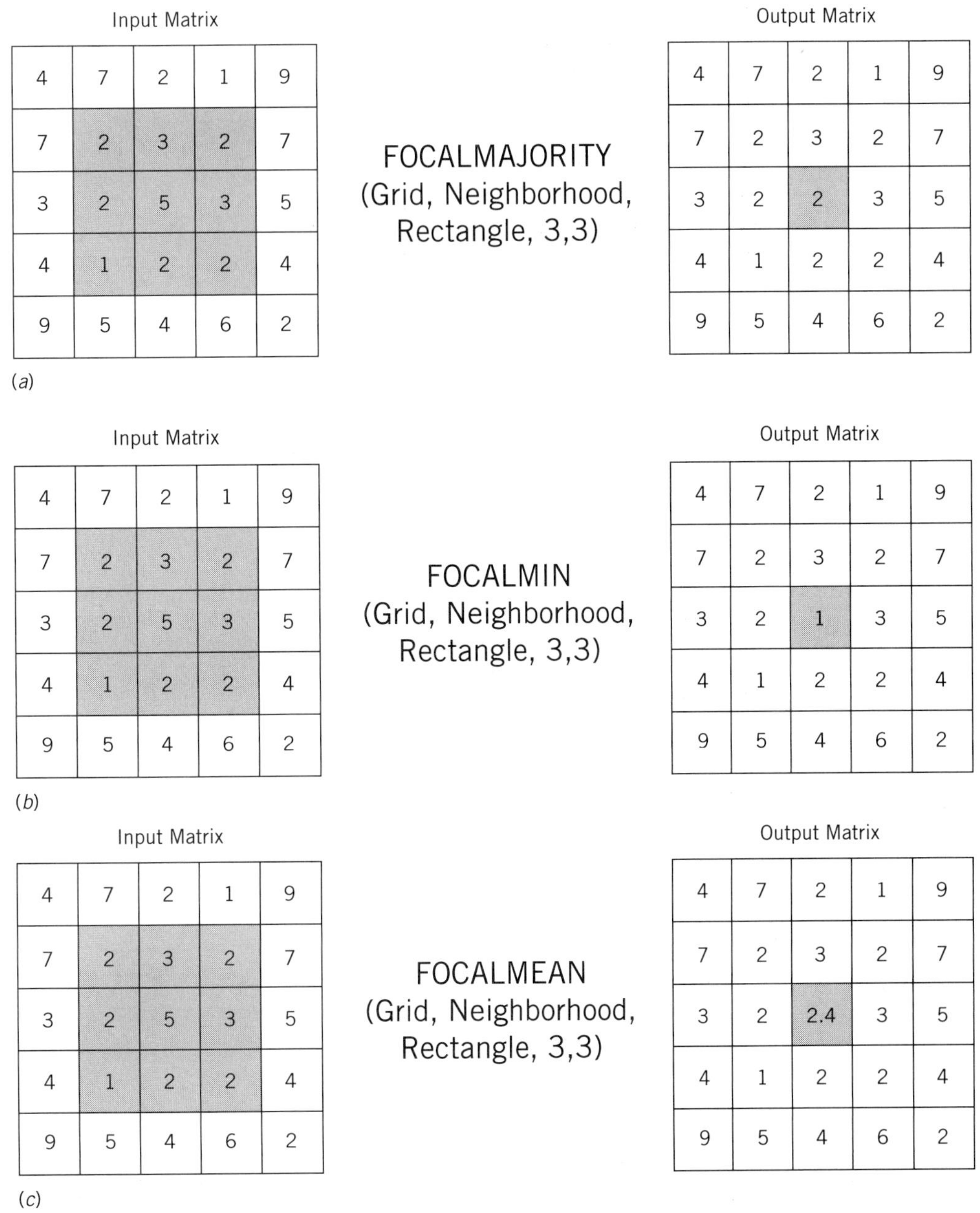

Input Matrix

4	7	2	1	9
7	2	3	2	7
3	2	5	3	5
4	1	2	2	4
9	5	4	6	2

FOCALMAJORITY
(Grid, Neighborhood,
Rectangle, 3,3)

Output Matrix

4	7	2	1	9
7	2	3	2	7
3	2	2	3	5
4	1	2	2	4
9	5	4	6	2

(a)

Input Matrix

4	7	2	1	9
7	2	3	2	7
3	2	5	3	5
4	1	2	2	4
9	5	4	6	2

FOCALMIN
(Grid, Neighborhood,
Rectangle, 3,3)

Output Matrix

4	7	2	1	9
7	2	3	2	7
3	2	1	3	5
4	1	2	2	4
9	5	4	6	2

(b)

Input Matrix

4	7	2	1	9
7	2	3	2	7
3	2	5	3	5
4	1	2	2	4
9	5	4	6	2

FOCALMEAN
(Grid, Neighborhood,
Rectangle, 3,3)

Output Matrix

4	7	2	1	9
7	2	3	2	7
3	2	2.4	3	5
4	1	2	2	4
9	5	4	6	2

(c)

Figure 4.10 Focal functions. Focal operations can contain a wide range of evaluators. Here we see how we can look at a 3 × 3 set of grid cells and evaluate them for their (a) majority value (b) minimum value, or (c) mean value. The result is returned to the central cell.

ment of plant propagules (seeds) blown by the wind through a hedge; the movement of animals across a patch in the landscape; the flow of water; or even the movement of ideas (a concept often referred to as innovation diffusion), fire, natural disturbance, or many other items across and through neighborhoods that differ from their surroundings. Some grid systems allow you to perform focal flow functions across the neighborhoods. Or you could create your own flow operation for each neighborhood. In ESRI's GRID software, this function is specifically defined to use an immediate 3 × 3 neighborhood and determines which of the central target cell's neighbors flow into it. Flows are assumed to move from higher numbers to lower numbers. One additional major difference between the focal flow operation and the other statistical

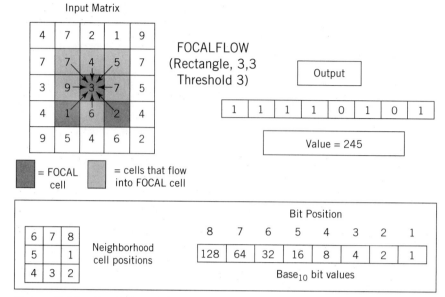

Figure 4.11 **Focal flow.** A 3×3 rectangle is evaluated relative to its target cell to decide whether each neighbor cell flows into the central cell. In this example, the cross-hatched cells flow toward the focal or target cell.

focal operations is that the output is a pattern of grid cells (again, 3×3) that shows the 8 grid cells surrounding the target cell whose values are higher than the target cell itself (Figure 4.11). The output from this operation is a bit map containing cell location information and 0 or 1 output depicting whether that particular cell flows toward the target. When all grid cells have finally been evaluated, the bitwise number is converted to base 10 and is then output. This is where the bitwise operators come in handy for using the output from this function within other functions and allowing the meaning of the operation be be maintained.

Zonal Functions

Zonal, or by-zone, functions have strong similarities to focal functions, especially in that both are based on the idea of characterizing a neighborhood (Martin 1996). So as with focal functions, they create output grids based on target cells within neighborhoods called zones (Figure 4.12). Although the zonal functions also operate on the idea of a neighborhood, the definition of the neighborhood (zones) is usually reserved for what is defined in geography as formal regions. Regions (zones) are defined in the raster GIS context as groups of grid cells that share the same values. Regions or zones can be contiguous (all connected cells), fragmented (grid cells are not connected), or punctate or perforated regions (regions with holes). With zonal functions, the zones are normally already defined in a separate grid, indicating that two input grids are normally required—the first that defines the zones and the second on which the statistical functions will operate (Figure 4.12). To keep our terminology consistent with that of Map Algebra and to remind us that the zones of grid cells are defined differently from the neighborhoods defined through focal functions, we will continue to refer to them here as zones.

Like focal functions, we can evaluate the neighborhood (region or zone) of grid cells with the following general or statistical operators: Min, Max, Majority, Mean, Median, Stdv, Variety, Range, Sum; there are many more. There are also a group of geometric operands that can output geometric measurements about the zone in a grid,

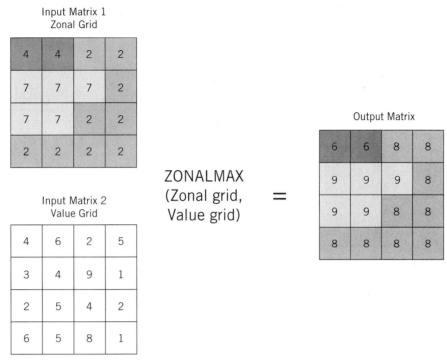

Figure 4.12   Zonal functions. Zonal functions evaluate those grid cells that are contained within the zone (or region), whether the zone is contiguous, fragmented, or perforated. The results are returned not to a target but rather to the same grid cells as the zone itself.

such as area, perimeter, and even thickness (the thickest point within each zone) (Figure 4.13). Whether statistical or geometric, the focal functions present values either as output grids where all of the grid cells for each zone contain the identical value or as output tabular data in the extended raster model. When output as a grid, each function used in the evaluation corresponds to a single type of statistic or value

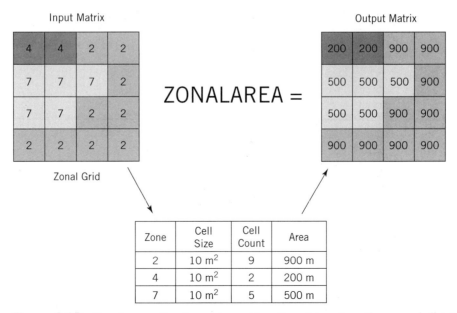

Figure 4.13   Zonal area. One type of zonal function determines the area of all grid cells contained within the zone. The total area is then returned to each grid cell within the zone.

feature and the result is assigned to each cell of the output grid. In the extended raster model, each type of output data is stored as a separate item in the database.

The geometric functions provide some very useful information for performing additional shape analysis routines, as one might encounter in landscape ecology and related disciplines (McGarigal and Marks 1994). Zonal area and zonal perimeter are fairly straighforward. Zonal area calculates the area by counting the number of grid cells and multiplying that by the area of each cell. It is important to remember that zonal area works on zones, not on isolated groups of cells. So if your zone is a fragmented region of, say, three fragments, the area will be the total area of all three fragments. Zonal perimeter sums the lengths of the interior and exterior sides of the cells that compose the zone. As with zonal area, the zonal perimeter is a function of all zone fragments in a fragmented region and the total perimeter in all perforations. For both perimeter and area calculations of a zone, if you need to isolate individual fragments you need to first isolate each by reassignment, reclassification, or selection so that each fragment is a unique zone or region.

Block Functions

Block functions are essentially modified versions of focal functions. They use a form of moving window, typically a rectangle, not unlike the focal function moving window filtering functions. The differences involve how the resulting values are saved and how the window is moved. In the case of block functions, all the values in the block are evaluated and all the corresponding grid cells in the output grid are assigned the output value. Once this is accomplished, the entire block is moved to a pristine area where calculations have not been employed (Figure 4.14). As with the focal opera-

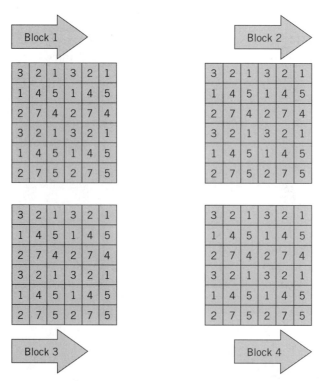

Figure 4.14 Block functions. Unlike focal or zonal functions, block functions operate on discrete blocks, one at a time, then move to another unique block. Such roving window functions are common in filtering algorithms within remote sensing software.

tions, the typical operations include some or all of the following: Min, Max, Majority, Mean, Median, Stdv, Variety, Range, Sum, … Figure 4.15 shows the results of four typical block functions, Min, Mean, Variety, and Sum. The important thing to notice from these graphics is that the results are stored as a complete grid the same size, shape, and location as its input. Once this is accomplished, the block moves to a completely new location where none of the grid cells that have been operated on are included. In other words, each time the grid is evaluated, it is totally unique. Consider how this differs from, for example, focal functions.

Global Functions

We have seen how we can operate on a cell-by-cell basis (local functions); we have seen how we can extend our view of our grid by using neighborhoods (with focal functions), contiguous regions, fragmented regions, perforated regions (with zonal functions), and unique rectangular sets of grid cells (with block functions). Now we move to what is truly a bird's-eye view by considering our entire grid at once using global functions. Evaluations from global operations include simple Euclidean as well as functional (cost) distance measurements and a host of others that create output that may be, but does not necessarily have to be, a function of all the cells in the entire grid. Because the output from global operations may be functionally related to every grid cell in one or more grids at any given time, it is essential that the software have access to all of these cells. Unlike the other functions we have seen, the groups of global functions are often radically different from one another. We can divide them into the following:

- Euclidean distance global functions

- Weighted distance global functions

- Surface global functions

- Hydrological global functions

- Groundwater global functions

- Multivariate global functions

Euclidean Distance Functions   Euclidean distance functions are designed to calculate distance measurements from a source cell or group (neighborhood) of source cells. It calculates both the distance (EucDistance) and the direction (EucDirection) from the source (whether an individual grid cell or group of grid cells) to its closest neighbor cells. Additionally, the EucAllocation function isolates which grid cells are allocated to which source cell or neighborhood on the basis of a determination of which are in closest proximity to the source. EucDistance is calculated from the center of the source to each of the surrounding grid cells. Some simpler raster GIS software packages use a cell distance whereas more sophisticated software packages use the maximum grid cell distance (based on the cell resolution), then uses the Pythagorean theorem to calculate the hypotenuse of the right triangle (Figure 4.16). In the latter case, if the calculation of shortest distance is less than some specified maximum distance, the value is assigned to the output grid cell. Also, if your software uses the Pythagorean theorem, the distance value output is a rational number. There will often be cases in which a source cell is at an equal distance from two or more targets; the value is assigned to the grid cell that is first encountered in the search. For this reason, you should be aware of the distance and direction of scan used by

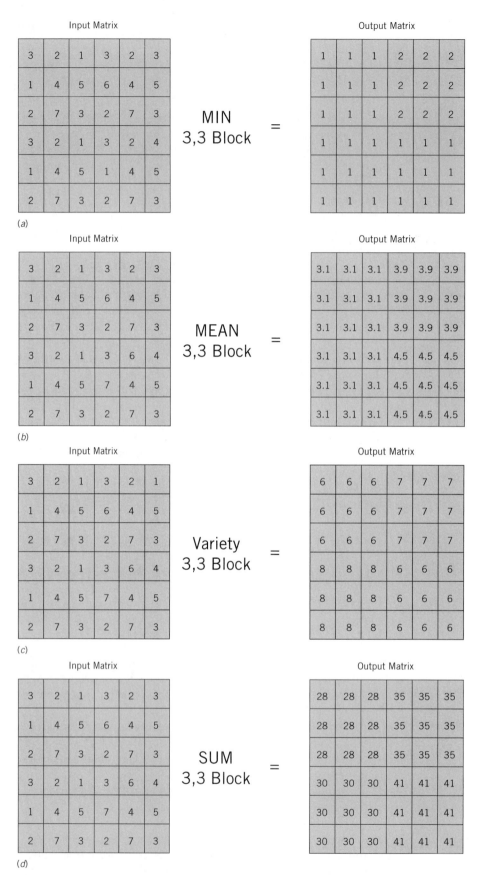

Figure 4.15 Sample block functions. Four ways to use the block function to evaluate 3×3 blocks of grid cells: The values returned are (*a*) the minimum, (*b*) the mean, (*c*) the variety (number of different numbers), and the (*d*) sum of all nine grid cells within each block.

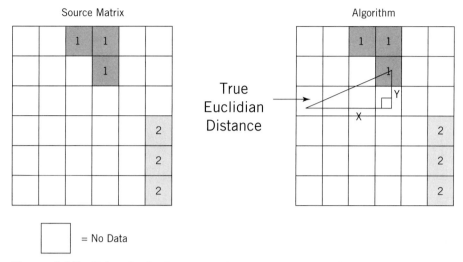

Figure 4.16   Using the Pythagorean theorem to evaluate the correct distance on a diagonal. This theorem is employed whenever grid cell distances are calculated on the diagonal to ensure accuracy.

your software. ESRI's GRID software, for example scans on a row-by-row basis, starting in the upper left of the grid.

Depending on your software, the EucDirection metric can assign code numbers representing the cardinal directions from the source cells, plus additional values for noncardinal vectors. For example, you could use a clockwise system in which you have 0 representing either 0° or 360°, 1 representing 45°, and so on. Or you could assign actual compass directions in the output grid cells, using a 360° compass (Figure 4.17). In this instance, 360° would be used to represent north, thus reserving the value 0 for source cells. These are only a couple of examples of systems that could be applied.

The EucAllocation function produces an output grid that records, for all grid cell locations, the identity of the closest source cell or neighborhood. To record the allocation, the grid values for each source cell or neighborhood is recorded for the allocated cells. So if you had two source cell neighborhoods numbered 1 and 2, those

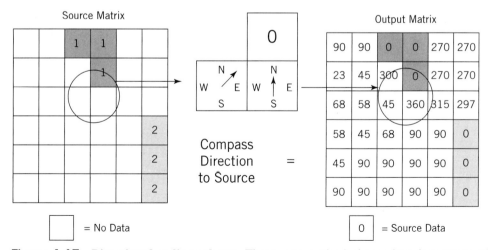

Figure 4.17   Directional coding scheme. This is one method of encoding directions within a grid-based geographic information system. Here, we see 360° compass directions being employed.

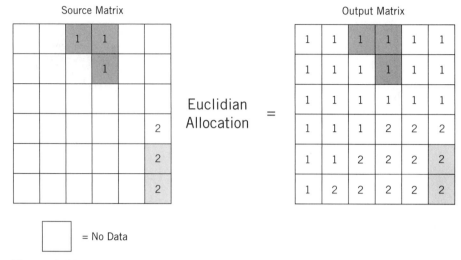

= No Data

Figure 4.18 Euclidean allocation. This function records the location of the nearest grid cell to a set of unallocated grid cells. The output records the number already assigned to the nearest grid cell values.

cells allocated as being closest to source cell 1 would be assigned a value of 1 and those closest to source cell would be assigned a value of 2 (Figure 4.18).

Weighted Distance Functions Weighted distance functions are based on the concept of accumulated travel cost distance from each cell to the source cells. The accumulated cost distance can also incorporate the idea of a friction surface and therefore produce output cells that more closely approximate the concept of functional, rather than Euclidean, distance. The costs of travel (functional distance) may be based on travel time, or they may be associated with monetary expense (e.g., taxi costs or cost of gas), or they might be based on some function of aggravation or preference. To perform these calculations, the software requires both a source grid and a cost or friction surface or impedance grid as input. The source grid can contain a single cell, multiple cells, or single or multiple groups of cells. Because of the way in which distance is calculated, the software evaluates only grid cells with a value of 0 (for starting cells) or greater for subsequent cells. If you wish to avoid evaluating selected grid cells, they must contain no data.

There are several ways in which raster GIS software can perform the cost distance operation. All of them combine an additive function of EucDistance combined with the friction or impedance values assigned to a friction or impedance value—sometimes called a cost surface. If you are traveling from cell to cell along horizontal or vertical directions, you add each grid cell cost and divide by 2 on the basis of the following formula:

$$a1 = \frac{cost1 + cost2}{2}$$

where $a1$ is the cost for each pairwise cost, $cost1$ is the assigned cost for the first grid cell, and $cost2$ is the assigned cost for the second grid cell encountered. This is then easily generalized for movement from the second grid cell to the third by replacing the cost values thusly:

$$a1 = \frac{cost2 + cost3}{2}$$

where the new value, *cost3*, is the assigned cost for the third grid cell. In this case (the horizontal or vertical case), the accumulated cost distance is calculated by simply adding each of the link values to achieve the following equation:

$$accumulated\ cost = a1 + a2$$

Of course, movement in a grid is not restricted to horizontal and vertical directions. Diagonal movement requires a distance of travel greater than a single grid cell length, and, based on the Pythagorean theorem for calculating the length of an hypotenuse, produces a multiplier of approximately 1.414216 (the square root of 2) when the grid cell length is 1.0 on a side. So the diagonal movement between the first two grid cells would be

$$a1 = 1.414216\ \frac{cost1 + cost2}{2}$$

And when accumulative cost for the diagonal movement from grid cell 1 to grid cell 2 to grid cell 3 is calculated, it uses the equation

$$accumulated\ cost = a1 + 1.414216\ \frac{cost2 + cost3}{2}$$

which, when simplified, looks identical to the formula for horizontal and vertical movement. The only difference, then, is the initial calculation of the link distances (i.e., *a1*, *a2*, *a3*, etc.).

The overall process of accumulated cost distances is an iterative process that begins at the source cells, then selects the lowest cost cell in the grid and accumulates values for the output grid in the following manner. First, the source grid cells are selected and assigned the value of 0, indicating that there has been no accumulation. Next, all the software activates all the cells neighboring the source cells, calculates their cost values on the basis of the previous formulas, and then selects which of these cells will be sent to the output grid. Cells assigned to the output grid must indicate the next "least-cost" path to a source.

Once the software selects an output (least-cost) cell, it adds this to a list of accumulative cost cells, then examines its neighborhood cells and determines which of these are capable of reaching a source. Only cells that have this ability are included in the list. As before, the cost of movement is calculated using the formulas above. The process continues by selecting the lowest cost, expanding the neighborhood around the chosen cells, and calculating the new costs and adding them to the active list. The process continues until the software encounters the edge of the grid, a window boundary, or the maximum distance (often preselected by the user).

A modification of the cost distance function is often called a drain function or path distance and is often used in conjunction with cost functions, EucDistances, hydrological functions, and other raster GIS functions to model dispersion and movement processes. The path distance function goes beyond the accumulation of a cost over a surface by compensating for actual surface distance traveled rather than a simple planimetric view of the surface, and it incorporates both horizontal and vertical components that impact movement from one place to another. This function is used for dispersion modeling and flow movement as well as least-cost path determination (a function often relegated to network modeling capabilities of a vector GIS).

A common example of the impact of one of these components is the impact of elevation on the fuel consumption of an automobile traveling from point A to point B. The total distance traveled, what we will call surface distance, and therefore the amount of fuel used are functions of both the planimetric (horizontal) distance

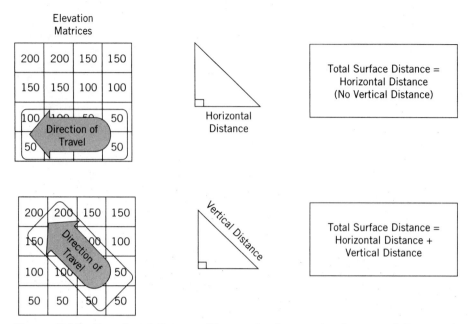

Figure 4.19 Functional distance. Distance is often not just horizontal. Frequently, additional components have an impact on the true distance traveled. In this example, we add the vertical component resulting from changes in elevation.

between the starting point and the destination and the vertical distance. The addition of vertical distance between the two points adds to the total surface distance traveled and results in greater fuel consumption. A simplified view of this is to employ the Pythagorean theorem to show the increase in travel on the basis of the differences in elevation between point A and point B (Figure 4.19). If, for example, the planimetric distance between the two locations is 25 miles and the vehicle gets 25 miles per gallon on flat surfaces, the trip would be expected to cost 1 gallon of fuel. However, if there is an elevation difference of 3 miles between the two points, we employ the Pythagorean theorem to obtain the total surface distance (sometimes called road log distance), and we see that the surface distance value has increased to approximately 25.18 miles. This means that we will consume approximately 0.18 gallons of additional fuel. Of course, this assumes three things—that the roads are perfectly smooth, that there is no wind, and that there is no impact due to gravitational attraction—assumptions that cannot be made if our calculations are to be at least representative of reality.

Surface roughness may not seem like a major component of path distance, but you might want to consider the impact of gravel roads versus highways, or pristine highways versus highways filled with potholes, on the speed at which you feel comfortable driving. Over long distances, even minor road surface roughness factors can have a compounding impact on the fuel consumption of a vehicle. A friction factor, as we saw in our cost distance function, can be assigned to compensate for these road hazard properties so that they can be incorporated into our path distance evaluation.

Our second assumption concerning wind conditions is one of a group of factors called horizontal factors that influence the travel cost. Motorists are aware that a wind coming from behind (tailwind) will often propel the automobile forward (Figure 4.20a) while reducing the fuel consumption, whereas wind moving into the vehicle (headwind) will increase the vehicle's fuel consumption (Figure 4.20b) because of the increased drag. Crosswinds, arriving at a variety of angles, have both positive and negative components, where the resultant vector is some combination of the speed and direction of the vehicle and the speed and direction of the wind (Figure 4.20c).

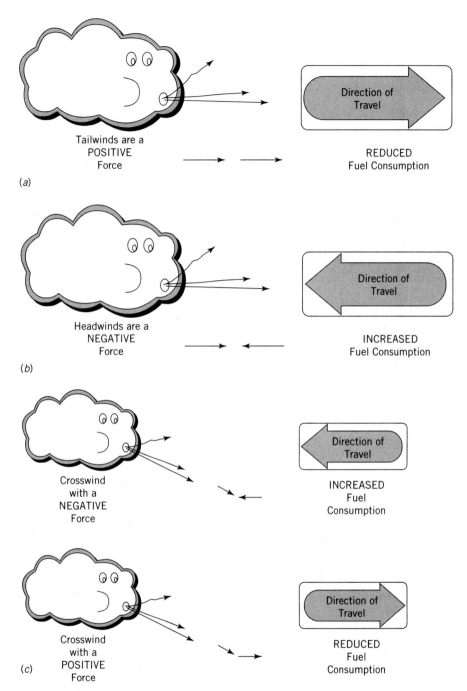

Figure 4.20 Horizontal travel cost factors. Wind resistance is an excellent example of how horizontal factors can affect functional distance. In this example, we see that (*a*) tailwinds reduce fuel consumption (our measure of functional distance), (*b*) headwinds increase fuel consumption, and (*c*) crosswinds have an effect based on the vector force and direction components.

The impact of elevation and gravitation on fuel consumption is a vertical factor and impacts the vehicle's fuel in one of two different ways, depending on whether it is traveling up- or downhill and, of course, on the slope of the hill associated with the vertical direction of travel. As you might guess, if you are traveling downhill, the fuel consumption is going to be reduced on the basis of the slope, whereas if you are traveling uphill, the fuel consumption will be increased, again dependent on slope.

A fully functional raster GIS should allow incorporation of all of these factors when evaluating path distance, whether it is modeling a source of dispersion or a moving object like a vehicle. One simple equation for the incorporation of these factors would be the following:

$$fuel\ used = SD \cdot F \cdot HF \cdot VF$$

where SD is the total slope distance, F is the surface friction factor, HF is the horizontal factor (in our case, wind resistance), and VF is the vertical factor relating to up or down slope movement. Each of the factors will have to be quantified and/or evaluated prior to being included in the equation. For example, the vertical factor will have to be evaluated first to determine slope (average slope between points A and B, in the simplest case). Additionally, we will have to indicate somehow whether the vehicle is traveling up or down slope.

Even this simple vehicular fuel consumption has some nasty details that make the model much more complex than has been enumerated here. Just a few you might have already considered are based on variability of each of the factors. It is unlikely, for example, that the road surface, wind direction and speed, surface distance, and slope will be uniform along the entire distance between points A and B. If our travel example were to be based not on vehicles but on biological phenomena or such physical phenomena as chemical spills or fire, many of our rules are either radically different or possibly even unimportant. For example, a nearly vertical slope filled with dry vegetation acting as fuel is likely to increase the movement rather than slow it down. In this same example, a rough surface comprised of, for example, dry woody shrubs, would be very difficult for many vehicles but would provide very rapid movement of fire. Another example indicates how the vertical component of your model may not even be a topographic slope. For such movement as that of hazardous gases from spills or sulfur-rich volcanic ash, the slope may actually be more a function of differences in barometric pressure resulting from differential heating and cooling than of topographic effects. Or orographic lifting might affect these barometric pressure values, thus complicating your model further.

As you can see, then, the modeling of path distance, as with cost distance, is unique to the phenomena being modeled and the environmental factors that impact them. A raster GIS capable of performing these global functions should also provide a flexible environment in which to control these factors. The actual modeling will most often be performed as an iterative process, nearly identical to that described for cost distance but with the inclusion of some or all of the factors just described. Rather than reiterate this, I'd like to take at least a quick look at the idea of friction coverages used in many global functions, especially those we have just examined.

Conceptually the idea of friction values seems quite simple: higher friction, higher friction values. However, as we saw in cost surfaces and cost distances, there are many forms of cost and many forms of associated friction that can be applied. Assigning friction values has a profound effect on the nature, reality, and acceptability of the models that employ them. Before you assign friction values, be sure to decide what level of geographic data measurement you need to employ (ordinal, interval, or ratio). This requires that you know the nature of the friction value (what it is meant to represent), whether it can be or has been measured, and how the values are meant to interact with any other operators, functions, or procedures while modeling takes place. For example, if you don't have actual friction values available through actual measurement, you might want to use some ordinal ranking values. Typical ordinal rankings may range from 0 (no friction) to 10 (maximum friction). This makes the assignment of friction values relatively simple. However, it also presents three major problems. First, the derivation of ordinal rankings is often somewhat arbitrary. Little research has been done on this subject, and what little has been done (Robinson 1990) has not been formalized within existing GIS. Second, ordinal values can be compared only within the spectrum from which they are derived, just as calculus grades can be compared only with other calculus grades, not with English grades. Finally, because of the second limitation, any mathematical manipulation of the rankings, such as might be done by Map Algebra multiplication by another coverage, is likely to produce invalid numerical results. Unfortunately, most raster GIS software is insensitive to geographic data measurement scales, allowing even numerical representations of nominal categories to be used within mathematical equations. In short, beware the source, measurement level, and use that you intend to make of friction surface values.

Surface Functions Any global functions would be incomplete if they did not include manipulation of surface features. Statistical surface data can be represented as integer data, with each grid cell representing a single whole number, elevational value, or rational data, where each grid cell is represented by a single floating point value. In most cases, these values are considered to be point values and are also most often stored as the centroids of each grid cell. Among the most common uses of these point data are for the generation of descriptive surfaces that allow prediction of future values and for the generation of surface-derived grid output.

The surface-generation algorithms generally include inverse distance-weighted interpolation, kriging, spline, and trend surface modeling, all of which predict output values based on a sample of surface point data. You should note that linear interpolation is not covered because the likelihood of its producing an accurate surface is low. Which of the remaining methods is used is a function of the type of surface you are trying to generate, the type of surface you are trying to model, and the distribution of sample points. Details of these methods are available elsewhere (DeMers 2000a), but for modeling it is useful to examine some of the more relevant properties and general applicability.

Inverse distance-weighted interpolation examines the linear distance between sample points and weights the interpolation value as an inverse of their distance. The idea is that those values that are nearest each other are likely to be more spatially autocorrelated and should provide more representative values than those that are at greater distance. In fact, to select this method, it is usually assumed that the locational variable (e.g., topography) should be locationally dependent (spatially autocorrelated). There are several options available for inverse distance weighting, depending on your software, and they include shifting from local to global control by modifying the power function. Larger power values result in lower influence from surrounding points, and the surface will be less smooth. If you are looking for more detail, this is useful. Other options inlcude controlling the numbers, locations, and methods of selecting the control points for interpolation (Hodgson and Gaile 1999, Philip and Watson 1982, Watson and Philip 1985).

Kriging is based on the idea of regionalized variable theory, which assumes that the spatial variability of the statistical values in the surface are statistically homogeneous throughout. Each of the number of kriging methods uses a mathematical function that models the spatial variations in Z values within the sample of points. A semivariogram is used to record and evaluate the relationship between the distance between points and the variation in Z values. There are many forms of kriging, but they can generally be grouped into spherical, circular, exponential, Gaussian, and linear methods. If there is an assumption that the spatial variation in statistical surface data contains some local trends or structural component, a general set of kriging methods called universal kriging is recommended. Universal kriging assumes that three components are operating at once—a structural component (drift) representing the overall shape of the surface, a random but spatially correlated component (e.g., surface roughness), and random noise. Once the structural component

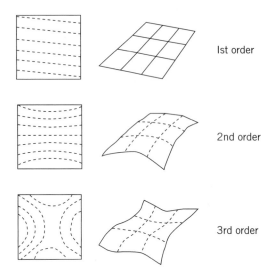

1st order

2nd order

3rd order

Figure 4.21   Trend surfaces. The more complex the polynomial equation designed to determine the general trend in a statistical surface, the more complex the trend surface itself. Here, we see the change from first- to second- to third-order polynomials and the impact this has on the shape of the trend surface created.

in universal kriging has been accounted for, the remaining variation is simply a function of distance, as with ordinary kriging.

Trend surface analysis uses a polynomial regression equation to fit a least-squares surface to the sample points. The purpose of trend surface is to show general changes in Z surface rather than to predict actual values from place to place. As the polynomial becomes more complex, so does the surface generated by the trend surface model. So a first-order polynomial creates a least-squares fit to a plane through the sample points (Figure 4.21). Second-order, third-order, and more complex polynomials create a least-squares fit that is more complex (Figure 4.21). Values beyond 3 really defeat the purpose of the trend surface and become too complex to be of use in predicting overall trends. Because the surfaces are more generalized than other methods of surface generation and because their primary purpose is to perform a best fit to the entire surface, trend surfaces seldom pass through the actual sample values.

Slope is defined as the amount rise (vertical dimension) over reach (horizontal dimension). Whether it calculates it as a degree value or a percentage slope (*rise/run* • 100), the output is a set of grid values that can be used for such operations as assisting us with our cost distance or cost path functions. Aspect, a feature that is essentially linked to slope, simply defines the direction of the slope by identifying the downslope direction of the maximum rate of change in grid cell values from each individual cell to its immediate neighbors. A good way of thinking about aspect is as the direction of the slope. Output values can be output as compass directions or as some compact coding values representing compass directions. Slope and aspect are almost always used in conjunction with one another when modeling.

Hill shading, sometimes called analytical hill shading, is a method of calculating the amount of solar illumination of a physical surface such as topography or buildings. Commonly, hill shading is thought of simply as a visualization tool, and it is certainly useful in this regard, as it enhances the visual appearance of the surface. It also shows areas that would normally be shaded or less exposed to visible inspection. As you might guess, the algorithms used for calculating slope and aspect are used for calculating a slope–aspect index that is used for creating shaded relief maps. For analysis, the hill shading algorithms available in high-end raster GISs allow for the analysis of the length of time and the intensity of sun at a given location. These values would be quite useful if you were, for example, analyzing your data for selecting appropriate sites for placing solar panels for generating electricity.

Hydrological Global Functions   Although we have seen how surfaces can be constructed and manipulated, there are some functional capabilities that are relatively unique when it comes to the modeling of water on the surface of the earth. Our primary interest is in the movement of water across the topographic surface, together with the movements of debris, pollutants, and biological materials. These types of functions are useful for hydrologists, planners, landscape ecologists, and a host of others interested in such movements for either theoretical or applied work. Examples include evaluating flooding potential and impacts, evaluating point and nonpoint source pollution loads to streams, and predicting the impacts of large-scale construction projects such as dams on the resulting flow characteristics of streams. One primary determinant in the overland flow of water is the shape of the terrain, most often modeled using DEMs to provide this information. They allow us to construct models of watersheds and stream networks, all designed to show where the water will begin and where it will arrive.

In the previous sentence, I used the terms *watersheds* and *networks*, both of which collectively define a drainage system. The two components, then, are the area over which the water will move and the network through which the water will ultimately travel, respectively. The drainage basin or watershed is the first component—the area over which the water drains into the more concentrated channels of a stream network. Other terms that are often applied to this are *catchment* and *contributing area*; these, along with *drainage basin* and *watershed*, are defined as the total area permitting water to flow into a given outlet or pour point. This outlet or pour point is defined as the lowest point along the drainage basin. This is important to remember for us to be able to model it effectively. It is also important to remember that the areas dividing the stream channels, the divides or watershed boundaries, are necessarily higher in elevation than the streams (Figure 4.22). When we model flows within a watershed, we must remember that the water will often move as overland flow until it becomes channeled within the stream network.

The network can be likened to a tree whose trunk is the lowest point (outlet, in our case). There are many types of branching forms for stream networks, each with its own unique effects on flow characteristics. I suggest you consult a basic physical geography text or a fluvial geomorphology text for more details. For our graphic example, we will use a common type called a dendritic pattern because it more closely conforms to our tree analogy. The branches of the tree (stream channels) intersect or are linked to each other at junctions or nodes. Outside links are those that contain no additional tributaries, whereas inside links continue to branch (Figure 4.23).

The characteristics of a stream network and its accompanying watershed have specific impacts on the movement of water. If you remember back to our discussion

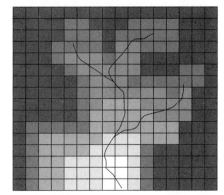

	Stream Channel
	0 – 50 ft elevation
	51 – 100 ft elevation
	101 – 150 ft elevation
	151 – 200 ft elevation
	201 – 250 ft elevation
	251 – 300 ft elevation

Figure 4.22   Grid representation of a stream's drainage. Note how the stream values are lower than the divide values between the stream tributaries.

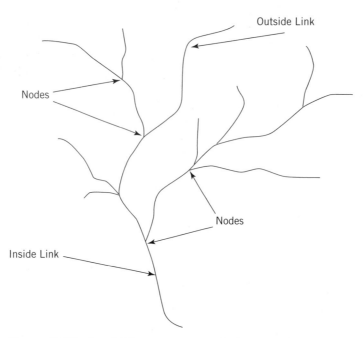

Figure 4.23 Dendritic stream network. Stream tributaries (links) are connected by nodes. Outside links contain no additional tributaries.

of slope and aspect within global surface functions, it is easy to transfer those basic ideas to surface hydrological modeling. The aspect, defined as the angular resultant from maximum rate of change in elevation from each cell to its immediate neighbor cells (slope direction), determines the direction of flow. The slope, defined as the maximum rate of change of elevation from each cell to its neighbor, will determine, to a large degree, the speed and energy of water flowing downhill. Steeper slopes result in greater stream energy and in a greater ability of the stream to erode and to transport sediment loads.

Because slopes are not uniform, either along the direction of flow or across it, we need to examine the impact of this slope variability on potential modeling activities. Profile curvature, the curvature or change in slope in the direction of slope, changes from concave to convex. Concave slope profiles, indicated by reduced slope along a portion of the stream, result in a reduction in stream speed and a concomitant decrease in energy. This, in turn, results in a reduction of the stream's ability to carry sediment load. As such, concave slope profiles result in either reduced erosion or increased deposition, or both. Modeling tasks requiring, for example, determination of areas of increased deposition resulting from damming streams will employ this basic principle. Conversely, convex slope profiles, typified by locally steepened slope, result in increased stream speed and power. This increased stream speed and power results in increased erosion and downcutting.

Surface curvature perpendicular to the slope direction is called planform curvature, and this indicates where the drainage basin profile or cross-section is either concave or convex. Concave cross-basin stream cross-sections result in flow convergence, or channelization of the flow. This normally results in rilling, gullying, and, eventually stream valley development. Convex cross-basin profiles are indicative of ridges and related divergent flow, whereby the flow is being directed to concave (downhill) portions of the basin.

Although not always the case, it is typical to use the cell-based DEM data for surface hydrological modeling, because of both their availability and because they are generally at a scale at which such modeling is most relevant. As described elsewhere

(DeMers 2000a), the DEM is a sampled grid or raster-based representation of continuous topographic surfaces on portions of the globe. Although many accept the true representation of topographic features using DEM data on faith, there are several things you should keep in mind before you begin modeling with them. First, the resolution of the DEM matrix will have a major impact on the resulting resolution in your modeling activities. But it is also not necessarily true that finer-resolution DEM data will necessarily give you improved spatial accuracy either. Finer-resolution DEM data are also more sensitive to other types of error that creep into the data as they are created. Three such error types include peaks, pits, and systematic errors. Peaks are unexplained high values or spikes that do not relate to the actual surface features the DEM is meant to represent (Figure 4.24a). Pits, or sinks, are the reverse of peaks in

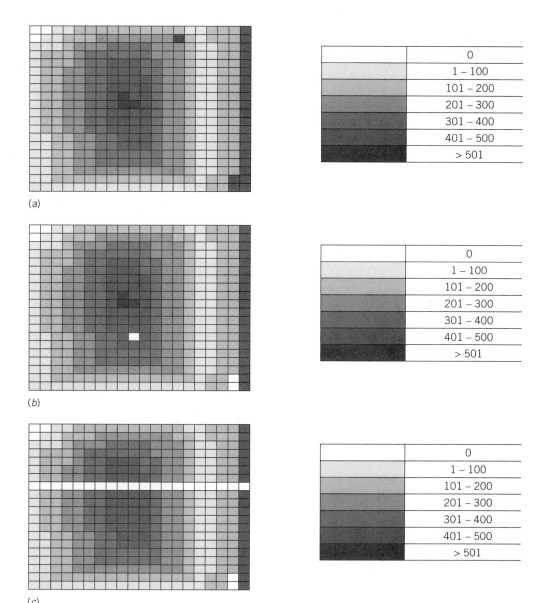

(a)

	0
	1 – 100
	101 – 200
	201 – 300
	301 – 400
	401 – 500
	> 501

(b)

	0
	1 – 100
	101 – 200
	201 – 300
	301 – 400
	401 – 500
	> 501

(c)

	0
	1 – 100
	101 – 200
	201 – 300
	301 – 400
	401 – 500
	> 501

Figure 4.24 Digital elevation model (DEM) errors. DEMs exhibit a number of errors of which the modeler should be aware. These include (a) unexpectedly high values, called peaks; (b) unexpectedly low values, called pits; and (c) systematic errors such as striping, where whole lines of data are either missing or obviously incorrect.

that they are unexplained low values that don't relate to surface conditions (Figure 4.24b). Some peaks and pits are, of course, true representations of land surface features, so it is important for you to know your study area before you begin working with these data. Some of these errors are minor and can be corrected in-house, whereas others may require that the data be returned and regenerated by the data provider. Systematic errors are often the easiest to pick up and are often characterized by a sudden shift of elevational values as you move across the map (sometimes called striping) (Figure 4.24c). These types of errors are most often indicated by an offset of the input device, are most noticeable on integer data within flat areas, and are unacceptable for inclusion in a GIS database. They should be returned to the data provider for correction or regeneration.

Now that we have examined the typical data set—the DEM—and have an understanding of the basic physical properties of watersheds, we can begin using our raster GIS to model the surface characteristics. Because the structural components of watersheds are primarily a function of slope and aspect, a determination of steepest downslope direction is our primary task. We can define our watershed boundaries, stream networks, and stream outlets once we know the direction of flow out of each cell.

There are several steps to using the basic functionality of the typical grid-based GIS to define watersheds and basins, to determine the accumulation of flow, and to model the length of flow within a watershed. Beginning with the basic DEM, we first evaluate the slope and aspect to determine the flow direction for the cells in our grid. Then we determine if sinks exist. If they do, they must be filled so they do not interfere with the overall flow-modeling process. The process of filling can be done using averaging procedures such as those employed by removing dropped point values in remote sensing (Jensen 2000). A simpler approach is simply to select the minimum elevation along the watershed boundary. You may remember we called this the pour point.

Once the sinks are filled, we have three basic types of function that can be applied to our grid. Accumulated flow is some form of cumulative weight of all grid cells flowing into each downslope cell in the grid. Watershed functions determine the contributing area (basin) to the overall flow. Finally, stream network functions evaluate the numbers of cells and the stream ordering of the overall stream network. We will examine the flow direction, accumulated flow, watershed, and stream network functions individually.

Flow direction is a key to much of the rest of the surface hydrological functions. This involves calculating the direction of flow for every grid cell in the entire grid. By using a surface (often a DEM) as the input grid, the software outputs a grid that indicates the direction of flow out of each cell. By starting at each individual grid cell, the software searches the eight surrounding grid cells and evaluates the direction of the maximum drop. This uses the following formula:

$$maximum\ drop = \frac{\Delta z}{d}$$

where Δz is the change in vertical value and d is the distance. As with our previous flow functions, the distance is typically measured between the centers of each grid cell. Between two orthogonal cells, the distance is 1; between two diagonal cells, it is 1.414216. If, within the immediate eight neighboring grid cells, the vertical drop is identical for all cells, the software expands the neighborhood until it finds the steepest slope. Once found, the direction of flow of the cell relative to its origin is coded with whatever value your software uses to indicate that direction. For example, you might use a coding scheme such as that used in ESRI's GRID software (Figure 4.25).

The following conditions that result during the evaluation of flow direction indicate sinks or areas of undefined flow direction:

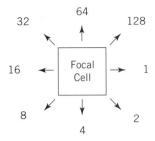

Figure 4.25   ESRI's directional coding scheme. The GRID software from ESRI uses a directional coding scheme for its global operations that moves clockwise from 1, which represents east. You should note that as you move clockwise, the values double. This helps to separate directional vector more easily than a simple 0-through-8 value method would.

1. If all neighbors are higher than the cell you are processing

2. If two cells flow into each other

3. If a cell has the same change in Z value in multiple directions.

For all of these conditions, the value for the output cell in the output flow direction grid is frequently calculated as the sum of those directions. For example, using the coding scheme indicated in figure 4.25, if the change in Z value is the same to both the right (coded here as *flow direction* = 1) and down (coded here as *flow direction* = 4), the flow direction results in 1 + 4, or a final value of 5. The most common method of eliminating sinks is to fill each with the value of the lowest boundary cell in the watershed. Some software has specific algorithms for doing this. Check your software user's manual to determine if these are available.

Again, as with other cumulative algorithms, we have seen that the flow accumulation is the accumulated weight of all of the cells that flow into each subsequent downslope cell in the output grid. Weights can be applied to the grid cells that might represent local additional input of precipitation or other factor contributing to the accumulation of flow within the watershed. If no such weights are assigned, a weight of 1 can be assigned to each grid cell, and the final accumulation will be the number of cells that flow into each successive cell. See Figure 4.26 for two graphic examples of a nonweighted (Figure 4.26*a*) and a weighted (Figure 4.26*b*) flow accumulation model. Whether weighted or not, the higher accumulation zones could easily be used to identify stream channels (Figure 4.27). Accumulation zones with zeros mean, of course, that no water is collecting, which would be indicative of ridges. The flow accumulation approach can also be used to model the amount of precipitation that falls on the upslope cells in a watershed or to represent the amount of rainfall that

0	0	0	0	0	0	0	0	0
0	1	1	1	1	1	1	3	0
0	3	1	1	1	1	5	1	0
0	1	5	1	1	7	1	1	0
0	1	1	7	9	1	1	1	0
0	1	1	1	16	1	1	1	0
0	1	1	1	18	1	1	1	0
0	1	1	1	20	1	1	1	0
0	1	1	1	24	1	1	1	0

(*a*)

0	0	0	0	0	0	0	0	0
0	2	2	2	2	2	2	6	0
0	6	2	2	2	2	10	2	0
0	2	10	2	2	14	2	2	0
0	2	2	14	18	2	2	2	0
0	2	2	2	32	2	2	2	0
0	2	2	2	36	2	2	2	0
0	2	2	2	40	2	2	2	0
0	2	2	2	48	2	2	2	0

(*b*)

Figure 4.26   Accumulated flow global function. Two methods of performing accumulated flow include (*a*) nonweighted and (*b*) weighted. Weighting methods allow for additional input of local precipitation.

Input Matrix   Output Matrix

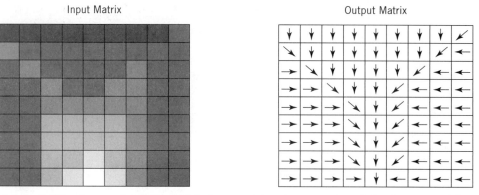

Figure 4.27 Accumulated flow. This is a graphic representation of the direction of flow resulting from the calculation of accumulated flow. Both weighted and unweighted methods make it relatively simple to isolate stream drainage patterns.

would flow from cell to cell. These uses, of course, would ignore water infiltration, evapotranspiration, and interception.

As indicated earlier, it is possible to take our definitions of what a watershed is and use our raster GIS modeling capabilities to define it in our grid. The delimitation of watersheds is a necessary element of nearly all surface hydrological modeling. Such boundaries can be combined with, for example, soil variables and land cover information to develop basinwide sediment loss or flood height models. This concept has been around for a while (cf. Band 1989a–c, 1993) and is now well established in many raster GIS software packages. We use a grid of flow direction as one major input to determine contributing area. Also needed is the lowest point in each watershed within a grid. You remember we called these pour points, and they should be able to be selected interactively. A background layer of flow accumulation could be effectively used to indicate the probable locations of the pour points. This will also ensure the selection of pour points for the major stream segments rather than for the minor tributaries. Selecting the minor tributaries will result in a definition of the watershed that is often substantially smaller than it is in reality.

In addition to the areal characteristics of the drainage basin itself, we should also have the capability not only to determine the locations of stream channels by using the flow accumulation, as we have already seen, but also to characterize the stream network itself (Jenson and Domingue 1988, Mark 1988, Tarboton et al. 1991). This is most effectively done by employing a simple thresholding limit value within a Map Algebra equation such that streams with a certain number of cells showing flow into them—say, 100 cells—will be indicative of streams within a network and will be assigned an output value of 1 whereas all remaining cells will be assigned a 0 value. These stream network definitions allow us to further characterize the individual stream components and to assign unique ID codes for components based on this characterization, and even to output the final results as a vector GIS coverage.

Among the most common approaches to characterizing stream networks is a technique called stream ordering that ranks the individual components of the stream network. Stream ordering is a method of assigning numerical values to streams on the basis of their position in the network. This is done by defining the number of tributaries branching from each trunk stream. Information from these orderings can often be inferred. For example, low-order streams tend to be dominated by more overland flow and less concentrated flow upstream. In fact, a first-order stream (streams that have no tributaries) have nearly all overland flow with no upstream concentrated flow. A direct result of this condition is that these streams are more likely to be susceptible to nonpoint source pollution, as one might expect from overland flow of

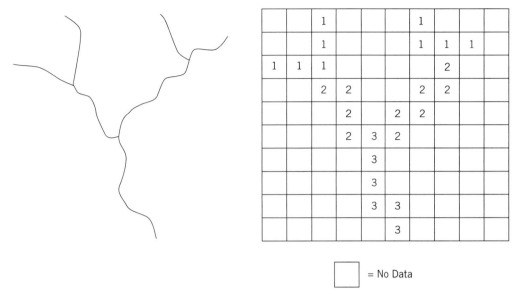

= No Data

Figure 4.28   The Strahler method of calculating bifurcation ratio in a raster geographic information system.

agrochemicals. Such streams would benefit from riparian (stream-based) buffers than higher-stream-order channels (Environmental Systems Research Institute 1994).

There are several methods of ordering streams, each of which ranks the branching streams differently. The most common methods are the Strahler (1957) method and the Shreve (1966) method. Most other stream ordering methods are derivatives of these two. Both of these general types of stream orderings begin by assigning an order of 1 to all exterior links. Stated differently, such links are considered first-order streams and are assigned the value of 1 as an ID code for first-order streams. The Strahler method, probably the most common method used, increases the stream order as each link is encountered. For example, the intersection of two first-order streams (two streams without tributaries) creates a second-order link (assigned 2 as an ID code) and the intersection of two second-order streams results in a third-order link (assigned 3 as an ID code); this continues until the trunk stream is defined as the final link (Figure 4.28). A major shortcoming of the Strahler method is that it increases in order only at intersections of identical order. For example, the intersection of a first-order stream with a second-order stream does not result in a third-order link. Instead, the method maintains the highest order of the highest-ordered link. As a result, the Strahler method does not always account for all links present in the drainage net, and it can also be very sensitive to the addition or subtraction of links to the net.

The alternative (Shreve) method does account for all network links. As mentioned earlier, this method, like the Strahler method, assigns a first-order categorization to all exterior links. Unlike the Strahler method, however, this method appends the links in an additive fashion. For example, the intersection of two first-order streams creates a second-order stream (1 + 1), the intersection of a first- and a second-order stream results in a third-order stream (1 + 2), and intersecting a second- and a third-order pair of segments creates a fifth-order stream (2 + 3) (Figure 4.29). In examining Figure 4.29, you will notice that the highest link number is 7. This is the number of upstream links— that is, the number of streams linking to the trunk stream. For this reason, the numbers associated with the Shreve method are called magnitudes rather than order numbers.

Groundwater Global Functions   Perhaps some of the most complex modeling capabilities of raster GIS are related to advection and dispersion modeling. Some software provides a specific set of algorithms for these capabilities. Rather than go

	1			1	1				1				1
	1	1		1					1			1	1
		1	2	1					1	1	2	1	
			2	2						2	2		
1	1		2	2					2	2		1	1
	1	1	1		2			2	2			1	
		1	1	3				2			2	1	
			3	5	2	2				2	1	1
				5					2	2		
				5	5	2	2	2				
					7							
					7	7						
					7							
					7							

☐ = No Data

Figure 4.29   The Shreve method of calculating the bifurcation ratio in a raster geographic information system.

through a rather lengthy discussion of all the equations, nuances, and details of groundwater functions, I will provide a very simple introduction and suggest a few references that can be applied to it. A complete discussion of advective dispersion modeling is available in Tauxe (1994). This approach employs three basic components. The first, Darcian flow, is based on the groundwater flow velocity and direction resulting from the transmissivity, head gradient (change in head per unit length in the direction of flow in an isotropic aquifer), and hydrological conductivity properties of geological formations. This is, of course, a simplification. A final result of these Darcian flow calculations is a grid known as a flow field. The second component models the path of movement through this flow field from some point source. This usually takes the form of a predictive model output based on the prediction of future locations of a hypothetical particle contained within the fluid. The process is based on an interpolation of nearest neighbor cells and is similar to the procedures found in Konikow and Bredehoeft (1978). Finally, the third component is based on the concept of Gaussian dispersion and relates primarily to the movement of dissolved material within the transporting fluid. This concept is based on two simultaneous mechanisms, advection and hydrodynamic dispersion. Hydrodynamic dispersion is the mixing of soluble materials (solute) with the solvent (transporting fluid) within the available pore spaces of the aquifer. Advection is the passive transport of the solute within the fluid. An assumption in the Gaussian dispersion component of the model is that the concentration of material is identical throughout the depth of the fluid.

Multivariate Global Functions   Multivariate global functions are not really functions per se. Rather, they are a subset of statistical techniques designed to explore complex relationships among many input variables, especially relationships that are not readily or easily observable. These statistical functions can be simple descriptive techniques like histogram and scattergram development, image/map compositing, and other methods of data display. Or they can be predictive and inferential statistical techniques like regression, cluster analysis, supervised and unsupervised classification techniques, and principal components analysis. Depending on your raster

GIS software, the multivariate statistical analysis function capabilities can range from prepackaged software programs and attendant data models, to more elemental procedures on which the more advanced statistical procedures can be constructed, to software with relatively limited multivariate capabilities but with links to packaged statistical programs. We will examine multivariate analysis in more detail later on; meanwhile, I'll provide a brief discussion, primarily to illustrate the potential modeling capabilities and to show the context in which such modeling exists.

First, we need to examine the difference between univariate and multivariate GIS techniques. A univariate technique might be an identification of regions as we discussed in zonal functions. For example, a typical application of such a univariate technique might be to identify all grid cells that share a common value (again a zonal function) such as all grid cells that share a common range of elevation values (slope). This region might be applied, for example, to indicate where particular crops might be effectively planted. By contrast, a multivariate technique might combine aspect (the direction of slope) to isolate not only regions that have a particularly useful slope value but also areas that might be facing the sun for improved plant growth. It might further include distances from roads, soil variables, zoning, and a host of others. One approach to such a multivariate model would be to go through a series of procedures to isolate all of the important factors, then perform some overlay operations. A more statistical technique might be to perform a cluster analysis that could effectively combine all of the factors in a statistically verifiable manner. The result would be a model showing the best locations for particular crops. This demonstrates the power of multivariate functions inside a GIS.

Here are a few examples showing how multivariate functions can be used to explore spatial relationships among grids, predict potential future conditions, perform terrain stratification, and even perform time series analyses. Suppose, for example, that you have a large sample of data on the number of interior birds located within a patchy area of a rain forest—say, an area of the rain forest that has large clear-cut patches. A linear regression model could be used to indicate the probable number of interior species in all patch sizes of rain forest contained within a grid. Of course, nonlinear models and multivariate regression models could also be applied to more complex situations.

Another form of predictive modeling based on the presence or absence of data, called a logistic regression model, has already proven useful for analyzing known locations of squirrels (Pereira and Duckstein 1993), comparing these locations with environmental attributes. This technique requires that you know the environmental conditions at a sample of locations where a particular animal, plant, or even a criminal is located. It also requires that you have a similarly sampled set of locations where the creature is not located. With these locations, you can predict the probability of finding each of these on the basis of the level to which each of the environmental conditions is met.

Remote sensing examples are perhaps the most common for illustrating the idea of the use of multivariate analysis for classification. A supervised classification, for example, requires the user to select a set of known values for, say land cover, as observed through multiband remotely sensed data. After selecting a number of grid cells (pixels, in remote sensing) that are known to represent the reflectance values characteristics of land cover types like water, row crops, forest, and urban, we can statistically compare these reflectance values against the unclassified pixels in the image. In this way, you can classify the remainder of the map, resulting in a map (an image map, in this case) of land cover.

Unsupervised classification techniques, again most often (but not exclusively) applied to remotely sensed data, can also be applied in a multivariate environment. By comparing relationships among, for example, slope, aspect, solar illumination, profile, and other variables, we can use the clustering algorithms to group the otherwise disparate data into logical groups or classifications. This is often done by randomly selecting seed values and, prior to knowing what they represent, trying to find similar groups of grid cell combinations. This is the statistical equivalent of a posterior aerial photo interpretation—the idea of creating classes of similar tone, texture, and so forth, then labeling them later.

GOING BEYOND MAP ALGEBRA

Although we have seen that Map Algebra brings with it an incredible amount of power, there are four additional lessons I want to leave you with. The first is that just as with any language (computer or otherwise), knowing all the words, what they mean, and the syntax for sentence construction is not enough to become a good modeler. Any number of books on Visual Basic or Visual C++ or any other computer language will provide a ready list of the commands and how they can be put together. However, just as knowing all the words in the English dictionary, their meanings, and their syntax does not mean you will be able to create a best-selling novel from them, knowing the language and syntax of Visual Basic and Visual C++ will not mean that you can create useful computer programs, especially complex ones. And, of course, the same is true of higher-level languages such as Map Algebra.

The second lesson is that, as with natural language, among the best ways to learn Map Algebra is by example. Later in this book, we will examine some of the published literature relating to GIS models created by others. We will examine the authors' approaches, designs, successes, and failures, and in some cases, we will even look at portions of their Map Algebra constructs. This will allow you to see how others have modeled within different settings. Hopefully, some of these will be similar to your own.

The third lesson is that like all other languages, Map Algebra has the capacity to grow. With the move toward object-oriented GISs, this capacity is going to become even more evident. Currently, most raster GIS software has its own programming language (some like Map Algebra and others not), and some have created a group of object-oriented modules (e.g., Map Objects) that will add to your tool kit and allow for more rapid expansion of the software, to simplify its user interface or its operations or to subset the package for smaller, user-specified needs where the enormous power of the GIS is really more than is needed for the task.

And finally, the fourth lesson is that as stated in Chapter 1, complex GIS models can nearly always be broken down into smaller, manageable parts. This is the subject of our next few chapters. We will begin by examining modeling essentials and modeling terminology, then proceed to conceptualizing, formulating, flowcharting, implementation, and finally, model verification. Model conceptualizing, formulation, and flowcharting are based entirely on the idea of breaking a larger model into its logical component parts.

Chapter Review

Functions bring together the more elemental operators within Map Algebra to solve a wide range of problem types and are categorized on the basis of their functionality. Local functions provide a worm's-eye view and operate on a cell-by-cell basis. This allows for reclassification and mathematical manipulation of individual grid cells one at a time or collectively within an entire grid. Focal functions employ most of the available operators within neighborhoods of grid cells. The neighborhoods can be either static or roving window clusters of contiguous cells. These neighborhoods can be selected in a wide range of definable shapes, from rectangles and triangles to circles, wedges, and doughnuts. This variability of shapes allows for the characterization of individual output cells or groups of cells on the basis of the attribute properties or distribution of properties found in the neighborhoods. Zonal functions employ many of the same operators, neighborhood shapes, and evaluation criteria as focal functions but offer the flexibility of characterizing output grid cells on the basis of the concept of the region. These regions can be contiguous cells, groups of cells with perforations, or disconnected clusters of cells all sharing the same characteristics. Block functions are similar to roving window focal functions, with the difference that after each evaluation, the window is moved to a unique group of cells before additional evaluation continues.

Global functions, unlike all of the previous types of functions, operate on the entire grid at once. These powerful functions can be grouped into the following categories: Euclidean and weighted-distance determinations; surface characterization and analysis; and hydrological, groundwater, and multivariate functions. Global functions allow movement through some or all grid cells within a single or multiple map. They are most often used when movement, dispersion, surface and volumetric analysis, and multicriteria evaluation require comparison of both lateral and vertical grid cell conditions.

Modeling with the functional capabilities of Map Algebra requires the recognition and use of four basic principles. First, an in-depth knowledge of the functional capabilities of your particular GIS software and its implementation of Map Algebra is essential to any successful model building. Second, among the fastest and most effective ways to learn how these functions are applied to your particular knowledge domain is to examine successful examples of their implementation. Third, like all computer languages, Map Algebra has the ability to grow through the development of additional algorithmic capabilities, created with traditional and object-oriented programming languages, to push the software beyond its existing limits. Finally, although GIS models can often appear to be impossibly complex, virtually all can be broken down into relatively simple modules, each of which can be built independently of the overall model, then linked for final model implementation.

Discussion Topics

1. Explain the potential problems associated with reclassifications of grid data when the result changes the geographic scale of data measurement. Include in your discussion the problems associated with local functional comparisons of one set of grid data to another.

2. Describe and provide a simple example of each of the six basic types of local functions (trigonometric, exponential and logarithmic, reclassification, selection, statistical, and arithmetic). Suggest how each of these might be used in a model. For example, how might one use the random within local functions?

3. Suppose you had a grid showing the depth to the top of an underground ore body recorded as individual depth values for each grid cell. Another grid shows the bottom of the ore body. Explain how you could create a simple equation using only local functions that would allow you to determine the best place to dig for ore. What other information would you need?

4. A common way of displaying aspect in raster grids is to use a number ranging from, for example, 0 to 359, where 0 could indicate that a particular cell is ori-

ented north. Discuss some alternative methods of encoding aspect. What advantages and disadvantages do compact methods have over the 360° method? Explain the two primary (noncompact) methods of depicting slope. How could local functions be used to isolate areas that had slopes of greater than 30° and were either south or southeastern in slope?

5. Some have said that slope and aspect cannot be separated from one another. Explain this statement.

6. Provide some concrete examples of how local functions can be used. Include in your examples such basic functionality as reclassification and map overlay.

7. What is the difference between attribute-based and locational-based selection procedures?

8. What is the fundamental difference between local and focal functions? Between focal and zonal functions?

9. In focal operations, the neighborhood is largely dependent on the position of the target cell. Describe and illustrate how the placement of rectangles, circles, wedges, and doughnuts are dictated by the target cell.

10. Describe what block functions are. How are they similar to focal functions? How are they different?

11. In focal and zonal functions, there are several optional shapes available for defining the neighborhoods. Provide some examples of when you might select such shapes as wedges, circles, and doughnuts. Hint: Think of the shapes of objects you encounter on a regular basis on the landscape.

12. Describe how such functions as focal majority, focal diversity, and focal minimum are used to characterize neighborhoods and then used to create an output grid cell.

13. Describe how global functions could be used to generate a surface that shows the relative difficulty of the movement of ancient hunter-gatherers across a terrain. Suggest at least two additional scenarios in which functional distance might be applied.

14. What is the difference between cost distance and path distance? What is surface distance? How does it differ from planimetric distance? Provide some examples of vertical and horizontal factors that might complicate path distance models.

15. What are the similarities and differences between the Strahler and the Shreve methods of stream ordering?

16. Suggest what types of functions would be used for a buffer, for a viewshed, and for the impact of a shelterbelt on erosion.

17. What is the difference between multivariate and univariate geographic information system analysis?

Learning Activities

1. Given the following grid cell value, perform several local functions (at least one from each of the six basic types). Two grids are provided for functions requiring comparisons (i.e., overlay operations).

3	3	1	4	8	1	0	0	0	0
3	2	0	5	6	4	3	0	0	0
3	4	2	2	9	3	3	3	0	0
0	1	1	0	2	5	4	3	3	0
7	1	1	1	2	8	1	3	0	0
0	1	1	1	1	2	1	1	1	0
7	7	1	1	1	1	2	1	1	1
6	1	1	1	1	2	1	1	1	0
1	1	1	1	2	1	1	0	0	0
1	1	1	1	2	1	0	0	0	0

Grid 1

3	3	2	4	8	1	0	0	0	0
3	2	0	5	6	4	3	0	0	0
3	4	2	2	9	3	3	3	0	0
0	0	0	0	2	5	4	3	3	0
7	0	1	0	2	8	0	3	0	0
0	0	0	1	0	2	0	0	0	0
3	7	0	1	1	0	2	1	0	0
7	1	1	1	1	2	1	0	0	0
1	1	1	1	2	1	1	0	0	0
1	1	1	1	0	1	0	0	0	0

Grid 2

For these grids, assign attributes for each operation you choose to perform. Whenever possible, describe what the operation means in terms of the changes in the attributes.

2. Using one of the two grids from question 1, select a target cell and a neighborhood for focal functions for each of the following shapes: rectangle, circle (approximate, of course), wedge, annulus. Illustrate what these shapes look like and where the target cells are.

3. For each of these shapes, perform a focal maximum, focal mean, and focal diversity operation.

4. Define three different zones (regions) from one of the grids in question 1. Include one contiguous zone, a perforated zone, and a fragmented zone.

5. Using a 5 × 5 block, perform a block maximum on each of the grids in question 1.

6. Again, using the first grid from question 1, assume that the numbers represent a set of codes for azimuth, where 0 represents north, 1 represents northeast, 2 represents east, and so on. The second grid represents slopes, where 0 represents flat, 1 represents 10°, 2 represents 20°, and so on. Now isolate all areas that face north, northeast, and east *and* that are greater than 10° in slope. What type of function is this?

7. Assume that grid 1 represents land cover. The number 0 represents abandoned grasslands, 1 represents fallow fields, 2 represents roads, 3 represents vegetable crops, 4 represents low-density housing, 5 is high-density housing, 6 is commercial land, 7 is industrial land, 8 is shopping centers, and 9 is the service and banking district. Use these nominal categories to define a new output grid that indicates potential friction values for movement across the grid.

8. On the basis of the friction surface you have created from question 7, perform an accumulated cost surface based on the interaction of the land cover grid and the friction surface.

9. Working with the grids defined in question 8, suggest how you would do a cost distance evaluation. Include an explanation of how you would incorporate vertical and horizonal factors as well.

10. Create your own simulated elevational grid based on the idea of a digital elevation model. Define how you would manipulate these values to identify streams and/or watersheds.

## CHAPTER 5

Modeling Essentials

On completing this chapter and combining its contents with outside readings, research, and hands-on experiences, the student should be able to do the following:

1. Using three or four examples from the published literature, isolate the spatial components that were used in creating the models and identify how these were defined as map elements for building the models

2. With those same articles, identify the ways in which the spatial components were linked with each other; these linkages may be logical, spatial, mathematical, heuristic, and so on, and may be explicit or implicit, real or inferred

3. Define methods for identifying spatial components and interactions for model building, including knowledge engineering, literature survey, and statistical techniques

4. Define and explain the following types of GIS (cartographic) models:
   - Descriptive
   - Prescriptive
   - Synthetic
   - Deconstructive
   - Empirical
   - Inductive
   - Deductive
   - Predictive

5. Explain the relationship between predictive modeling and descriptive and prescriptive GIS modeling

6. Explain the differences among atomistic, heuristic, and hybrid types of cartographic models

7. Discuss the advantages and disadvantages of using deductive logic in building cartographic models

8. Discuss the advantages and disadvantages of using inductive logic in building cartographic models

9. Discuss the role of descriptive modeling in producing prescriptive models

10. List possible solutions to the problem of building heuristics-based GIS models

11. Provide examples from the literature of GIS and cartographic models whose elements fit into the categories of models outlined in this chapter

THINKING SPATIALLY

GIS modeling is not an intuitive endeavor, especially for those whose thinking is not fundamentally grounded in geographic analysis and spatial thinking. Although the output of geographic analysis, particularly the cartographic output, leaves one with the misleading—in fact, often completely incorrect—impression that the process is simple and straightforward, quite the opposite is often the case. This is meant not to dissuade you from pursuing GIS modeling but rather to caution you against taking convenient, seemingly correct approaches when more complex solutions are necessary. Conversely, one should also not employ complex solutions when simple, more elegant ones are sufficient. Both of these problems occur when the power of the GIS is wielded without thorough preplanning and preparation.

Perhaps the best, first step prior to modeling is recognizing the spatial components of a problem to be modeled. Rather than beginning with the components themselves, it is often best to begin by examining the envisioned output from the analysis. Often called the spatial information product (Marble 1994), a recognition of the required output allows us to break the envisioned output and its model into interacting constituents. This is analogous to the process of dissection used in biology to begin to understand how the pieces of a subject fit together. And like the dissected specimen, the spatial information product allows us to make at least some basic estimation of the functional relationships of the components on the basis of where they are and how they are juxtaposed.

Each subject domain has its own unique data sets, interactions, and potential spatial information products. Still, there are some generalities that we can employ to begin our dissection. We can define these as a set of guidelines rather than as a set of strict rules. We will limit our discussion to cartographic spatial information products. Although there are many other types of potential output, most can be derived, in one form or another, from the map. The following guidelines should provide a basis for most modeling domains and scenarios:

- **Guideline 1:** A cartographic spatial information product (the output from analysis) contains within it a set of simpler, more elemental, intermediate cartographic output products.

- **Guideline 2:** The intermediate cartographic output products are most often simple models themselves, each of which is composed of cartographic or numerical elements that cannot be broken down.

- **Guideline 3:** The elements can be either grids (whole thematic maps), grid subsets (e.g., windows), numerical values, or variables.

- **Guideline 4:** The elements are connected to each other by operators. The operators represent the functional relationships among the other elements.

- **Guideline 5:** Some elements will be used more than once, both for intermediate model components. Some will be used more than once within a single intermediate model component.

- **Guideline 6:** GIS models are just that, models, and should present an opportunity for verification, validation, decision justification, and model refinement. All of these depend to some degree on an ability to explicitly define and explain all the components and interactions used in the model building process.

All of these guidelines are based on the premise that we are fundamentally using as input, analysis, and output, some cartographic representation of data representing selected aspects of the earth's surface—its physical, biological, and socioeconomic properties. In the present case, of course, we are limiting our representation and analysis (and often the output) to raster forms of cartographic representation. Still, the starting point for the development of nearly all GIS and cartographic models is the ability not only to define available cartographic data but also to have a fundamental understanding that there is a geography behind much of what we do.

It is this fundamental geography, and the common cartographic representation of this geography, that allows us to begin GIS modeling in the first place. DeMers (2000a) explained this as developing a functional "geographic filter." Some people seem to possess this ability, whereas others struggle. In fact, the geographic filter is readily available to all of us; it just seems to be more latent in some than in others. Robinson et al. (1995) stated that what we are doing is developing the art and science of graphicacy … the tendency or ability to think spatially about our world. Because we operate on a day-to-day basis within a spatial context, we frequently take the space or the geography for granted, operating nearly on automatic. When you lose your car keys, forget where you parked, or find yourself walking the wrong way to class, you have actually been distracted by other things than the familiar spatial environment. You didn't actually lose your keys, you simply placed them somewhere and failed to note the spatial locators that allow you to find them. The same is true of your parked car and your errant trip to the wrong classroom. Your spatial context is not explicit, is missing, or became confused by input of incorrect or incomplete spatial clues.

Although our movement and everyday experiences through geographic space are often recorded and synthesized to some greater or lesser degree by the analytical hemisphere (the left side) of our brain, a large part of our spatial cognition and understanding is found on the right side. Artists have found that it takes practice to reintroduce this hemisphere to its duties after long periods of inactivity (Edwards 1979). Because graphicacy requires both hemispheres to interact, we need even more practice than the artist might. We need first to be able to recognize that patterns exist. This is done by focusing our attention as we operate in our environment. We need to constantly be exposed to maps, aerial photography, and field excursions. We need to get up high and look down at our world—to sit in the window seat when we fly.

Once we begin to recognize patterns, we've take the first of two major steps in preparing ourselves to be graphicacy aficionados. Our next step is to begin to understand that patterns have associated with them some identifiable processes at work that created those patterns. After we have done this for a while, we will see spatial patterns everywhere, and we will immediately begin to ask why those particular patterns exist or why they change or how they might be important to our own knowledge domain. In fact, practicing GIS modelers are generally able to migrate from one knowledge domain to another and from one scale to another without a moment's hesitation.

With this advice provided, you will still require a kick-start, just as budding artists do when told that they need to exercise the right side of their brain. The next few paragraphs should provide that kick-start. I will use a number of examples that most of us are either familiar with or can become comfortable with in a very short time. As with the techniques in Betty Edwards's (1979) book, they are designed not as a replacement for the spatial experience but as some basic tools to help you develop the necessary skills. We will examine the isolation of both spatial patterns and spatial relationships—connected, but not identical concepts. We'll begin with patterns first and cover relationships next.

Let's begin by defining some basic terms. First of all, what does it really mean when we say we are going to isolate the spatial components of our problem? In aerial photographic interpretation, we often apply two related techniques when we evaluate a scene—identification and recognition. Identification most often implies that we already know that there are specific objects (or even patterns) contained within the image. Our task is merely to find them and tag their locations and is often considered the lowest level of feature extraction. This would be the equivalent of knowing that there are streams in a DEM (Figure 5.1). The purpose would then be to identify where, within the DEM, the streams are located in absolute terms. Although this information is useful, identification requires that we already know that some pattern or feature exists.

For our purpose, recognition is much more to the point. Recognizing spatial patterns is the first step in explaining and exploiting them for model building and is not unlike the first step in the scientific method—observation. For our discussion, we will begin by examining readily visible patterns on the landscape, then move on to patterns that are visible only if two or more themes are linked. We will then proceed to patterns that are more functional than immediately visible. In many of these cases, the patterns will either emerge through extended observations where the pattern is linked to changes through time or will require more complex methods of observation because either the patterns are obscured by seemingly nonpatterned landscape elements or the patterns are too subtle to be observed or are latent patternings requiring some forcing function to reveal their existence. Before we begin, let me remind you that the ability to observe patterns not only is a useful skill in GIS modeling but is absolutely essential. If no spatial patterns are perceived, there is little reason to

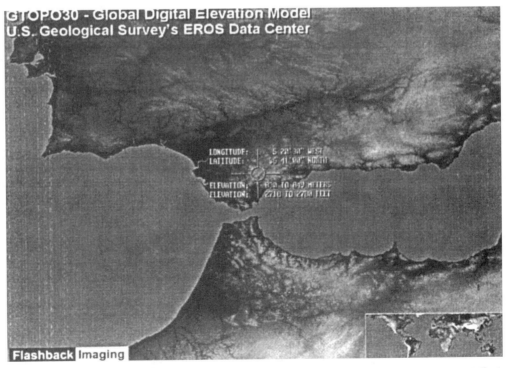

Figure 5.1   Digital elevation model with stream pattern. Note the linear arrangement that helps us define where the stream is.

perform GIS analysis. The next few paragraphs are designed to make you more sensitive to the types of spatial patterns that exist and to acquaint or reacquaint you with some of the methods and tools that can be employed to help you visualize them.

Visible Patterns

We begin our discussion of spatial pattern recognition with the most basic type, that of recognizing readily visible but often ignored spatial patterns. For earth and environmental scientists, the physical patterns on the earth are everywhere. For example, we can observe the random locations of individual plants and may hypothesize that these random distributions are a result of the method of seed and other propagule disseminations (Figure 5.2). In other cases, plants seem more clustered and seem more closely associated with substrate differences, or nonrandom methods of propagule dispersal (Figure 5.3). Still other observations may show uniform distributions of plants, most likely associated with some method of human intervention (Figure 5.4). Geomorphologists see land form distributions, such as alluvial fans resulting from the downstream deposition of sediment (Figure 5.5), dune patterns from selective aeolian processes (Figure 5.6), or glacial features such as boulder trains or glacial moraines whose striped patterns suggest the nonrandom processes of glacial retreat (Figure 5.7).

Landscape ecologists have taken the analysis of physical landscape pattern to new heights. Focusing on patches, corridors, and the surrounding matrix within which these features exist, they have created a robust set of mathematical descriptors to aid them in understanding the causal mechanisms of such patterns (Forman 1995). They have devised such landscape metrics as the number of patches in a

Figure 5.2   Photograph showing the random distribution of plants in a natural environment. Such random distributions are frequently a result of random seed dispersal mechanisms. Thus, a distributional pattern can be used to assist us in our geographic information system modeling by suggesting mechanisms related to patterns.

Figure 5.3   Photograph showing the clustering of trees. Again, we see how the distribution connects pattern with process, helping us in the geographic information system modeling process.

Figure 5.4   Regular distribution of trees in an orchard. Such distributional patterns again link pattern with process. Generally, regular distributional patterns are a direct result of human intervention.

Figure 5.5 Alluvial fans. Such geomorphological patterns as the alluvial fans found at the base of mountains provide us with insight into the movement of sediment that we might be interested in modeling inside a geographic information system.

matrix, the sizes and shapes of the patches, their between-patch distances and densities, isolation metrics, perimeter-to-area ratios, lengths and orientations, and many more (McGarigal and Marks 1994). All of these metrics are indicative of their increased sensitivity to the very existence of these landscape components, their

Figure 5.6 Dune patterns. Aeolian processes such as those observed in the patterning of dunes may be modeled in a geographic information system.

Figure 5.7   A portion of a map showing the regular series of arcs associated with the retreat of the glacier. Such features might prove useful in reconstructing glacial movement inside a geographic information system.

distributions, and other quantifiable descriptors—all related to an initial recognition of these patterns.

Urban and transportation specialists quickly recognize the differences between street patterns based on a formal grid versus those that are more closely linked to the underlying topographic, hydrological, or historical land subdivision patterns (Figure 5.8). The process of urban and regional zoning is an attempt to modify the land use patterns to control growth, improve access, provide patterns of green space, and provide for efficient use of dwindling resources. Once again, this process acknowledges the existence of some recognized existing patterns.

In all of these examples, the primary function of recognition involves some form of visualization of the patterns. Typical approaches to visualization include driving along road networks; making field observations of physical phenomena; and reading and analyzing aerial photographs, satellite images, and, of course, maps. The development of Mandelbrot's (1988) original ideas behind quantifying landscapes and other earth features was primarily driven by hundreds of airline flights in which he was able to observe over and over again the patterns of the earth from a high altitude. A long-held tradition among at least one group of geographers is to visit sites in the field and to photograph what they observe so they can document the

Figure 5.8 **Map of the city of Mankato, Minnesota.** Note how the regular grid pattern tends to break down near the river. Such patterns allow us to link physical features to planning.

observed patterns and recognized patterns they might have missed while at these sites.

These single patterns, and thousands more just like them, may require constant exposure on the part of professionals so that they begin to sort out the appropriate elements that create the patterns. Yet they are predominantly single-element patterns. By that, I mean that some single category of elements is applied to the identification of the patterns, such as forest patches in a grassland matrix (Figure 5.9), street patterns in an urban setting, or boulders on grassy terrain. Although such single-element patterns may be commonplace, there are many patterns that are less easily defined by a single category of elements and are, in many cases, harder to recognize. Just to assign a name to these, we will call these multiple patterns, not because there are several patterns involved but rather because the patterns are a result of more complex elements, often interacting (Figure 5.9).

Multiple patterns can, to the trained observer, seem just as elemental as single patterns. Or by contrast, elemental patterns may be viewed as too simplistic, needing to be further decomposed into more detailed elements. Once again, this is a function of experience, but more than that, the observer often possesses a more in-depth understanding of existing processes and therefore has a more refined intellectual filter within which to categorize the elements themselves. The complex elements are sometimes viewed as single categories, composed of unique mixes of primary elements. Or in other cases, untrained observers may view complex elements as single features, whereas trained observers see many more elements and interrelationships in the exact same landscape. For example, to an untrained observer, a corn crop may appear as a unique, single element in an agricultural landscape. To a trained agronomist, however, the field is often composed of many parts, each of which may indicate differences in pattern of plant maturity, which, in turn, may reflect differences in soil properties (Figure 5.10). In this way, two different pattern elements are observed at once, although they share approximately the same geographic space. Such spatially corresponding patterns essentially suggest a third pattern of interaction between the two variables. This is a fundamental relationship that is both a separate pattern in its own right and, more importantly, suggests functional linkages that will later be employed to create a model of, for example, crop yields related to some yet-to-be-discovered soils properties.

The example of the spatial correspondences of crops and soil suggests that the GIS itself may act appropriately to identify multiple-pattern correspondences. Using overlay functions within a framework of data exploration and data visualization

Figure 5.9   Aerial photograph showing forest patches in a grassland matrix. Such patterns are used by landscape ecologists to evaluate such function-related issues as the necessary size of patch for bird species and the effect of isolation on small mammals.

Figure 5.10   Aerial photograph of a cornfield showing variable maturity. Such patterns are often related to additional, often unseen, variables.

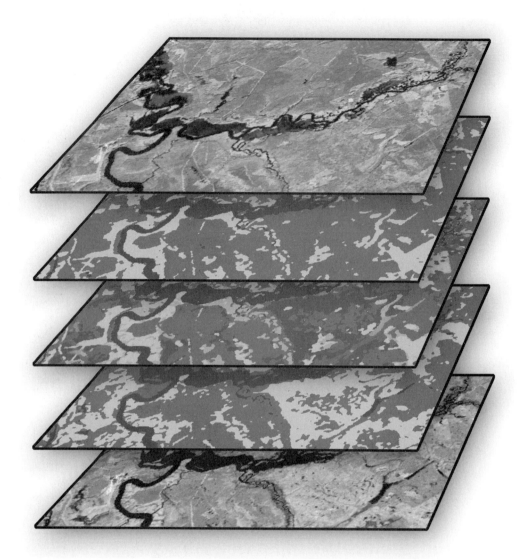

Figure 5.11   Overlay operations showing how one set of factors can spatially correspond with others.

allows discovery of correspondences of multiple variables (Figure 5.11). This not only is useful as part of the observation process but, as you will learn in more detail in this chapter, also is a primary function of descriptive cartographic modeling. We will therefore revisit the topic of multiple patterns later on in this chapter, both when we discuss functional patterns and when we consider the tools for examining patterns. The example also suggests that beyond there being just areal correspondences between or among variables, there is also some functional linkage that causes these correspondences.

Functional Patterns

Functional patterns may be observed by aerial correspondence, or they may actually be unobservable through such simple methodologies. This often suggests an even more in-depth knowledge of the spatial system being evaluated. One reason for a possible inability to observe functional patterns is that the patterns may be latent, requiring something to reach a certain functional or interacting threshold before visible patterns emerge. In our crop example from before, whatever soil properties may be accounting for differences in the crop itself may not appear until a certain stage in crop growth. Although certain soil properties have little impact on the emergence of crops because they are relying on food sources within the seed and on water and heat energy, other properties, such as certain micronutrients or soil texture, may impact the plants more as they emerge and begin to grow. Similar latent patterns often show up when chemicals that can negatively impact mature plants over time may have no effect on them until sufficient amounts are absorbed by the roots. In other words, a threshold value must be achieved before the patterns become visible in the plants, even though the harmful substances may already be present. Medical researchers have often found latent functional patterns of diseases such as cancer related to chronic exposure to hazardous materials (Harris 1997). Until sufficient intake of the materials over an extended period of time results in an identifiable increase in the number of cancer victims, the pattern remains unobserved. Although time is an element in such patternings, it is actually the threshold value of the substance and the reaction to it that causes the pattern to emerge.

Similar functional patterns may be related specifically to time with no reliance on thresholds at all. It seems that time, and our ability to observe time, is inextricably linked to the size and relative longevity of the observer. For example, insects, whose longevity is most often considerably shorter on average than that of humans, observe time at a completely different rate than we do. What we perceive as very rapid movements—say, a rapid shifting of our eyes—or swift arm movements may appear as very slow to an insect. This might explain why it is so difficult to catch a fly with your hands. The time differential accounts for our inability to observe many spatial patterns as well. Take, for example, the migration of animals within a landscape. To completely recognize what the patterns of movement might be, we need to use such devices as radio collars to follow the animals for weeks, or months, or even years before distinct spatial patterns emerge. The patterns are there, but they are perceived as latent patterns to us because of the time required to perceive them. Map overlay operations can also be used effectively to observe such temporal patterns as land cover change over years (Boerner et al. 1996), the process of plant succession, the clustering of criminal activity (Eck 1998), or differential traffic patterns at different times of the day or year. Even without the use of GIS overlay, it is still obvious that the proper sampling time frame is essential to allow recognition of the patterns. If you observe wild animals for only a 10-minute period, your observed patterns will be quite limited. Likewise, if you have never observed rush-hour traffic, you might assume that the flow of traffic in your city is always very fluid. And if you do not keep records of crime or land use and compare them after sufficient periods of time, the patterns will not be readily observable.

I have suggested some methods of observation that go beyond the simple eyeballing, such as the use of radiotelemetry and even GIS overlay. Like any other discipline, we need special tools to observe some patterns, especially latent, threshold-dependent, temporally unique, or functional patterns. Even what appear to be relatively simple visual patterns do not always become recognizable until we use different perspectives or different views. This was recently evidenced by some unique work performed by Peter Fisher (1995, 1996, 1998), whose examination of the common viewshed algorithms demonstrated that we had failed to observe what we really mean by a viewshed. He found, for example, that although we typically assume that if a house is on the viewer side of a hill it will be visible, the algorithm should reflect that. However, all other factors being equal, if the profile of a house extends above the hill, it is much more observable (much more obvious than one whose profile does not extend above the hill) (Figure 5.12). This could be extended to include

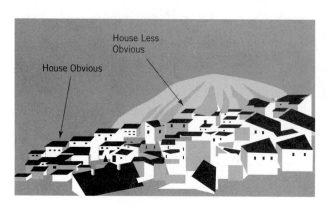

Figure 5.12 Buildings in front of a mountain but not extending above it are less obvious than those that do extend above it.

the important factor of contrast (often called the basis of seeing). The idea of camouflage is an attempt to confuse the viewer's visual perception by breaking up patterns so that they are not recognized. Although military camouflage does this on purpose, nature has a tendency to do the same as a matter of course. And just as we might need to use different sensing devices to recognize a camouflaged soldier, we need different tools to observe many otherwise obscure patterns.

In the case of our chemical spill example from before, our sight may need to be augmented by some form of soil sampling and mapping. The mapped distribution of the otherwise unseen chemicals will illustrate a pattern of chemical infusion even before the plants react to it. Many functional patterns are not visible until we link some samples to the use of such visual enhancement techniques as charts and graphs. A classic example of this is the use of line transects to sample temperatures through an urban area. When the temperatures are plotted, a concentration of heat, or an urban heat island, will appear within the core of the city (Figure 5.13). Similar belt transects will yield equally identifiable patterns of species concentrations, disjunct populations, or land values when subjected to simple graphing or mapping techniques.

An important—and often neglected—technique for identifying patterns is the use of statistical correlation and regression analysis techniques. Although many of us may be familiar with the techniques and even employ them in isolation, we often sep-

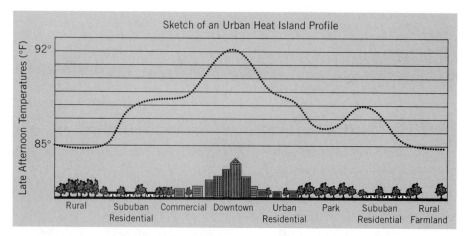

Figure 5.13 Cross-section of an urban heat island indicating that temperatures do demonstrate an identifiable pattern. *Source:* U.S. Environmental Protection Agency.

arate them from the process of GIS analysis. In fact, it is a good practice to employ such techniques for relating variables prior to using map overlay techniques. If we know in advance that there are statistically significant relationships among our mapped variables, we will be much more confident that the map overlay process both makes sense and is defensible during subsequent model verification. In addition, a highly useful technique is to map the residuals from regression to identify the variables that do not fit the model (Marble, 1965).

GIS modeling neophytes should be encouraged to examine their own domain knowledge through the use of both simple graphing and and more complex statistical techniques. It will prove gratifying to observe such patterns as the relationship between gas prices and nearness to major highways or the relationship between number of customers and distance to a store (whether actual or functional). To some, it may seem obvious that plants and animals seem to occupy coarse, continental-scale patterns, but it was not obvious to Charles Darwin and Alfred Russell Wallace until they sampled the animals and started making maps. The now classic 1854 map by Dr. John Snow showing the functional relationship to emerge between cholera and a nearby polluted well in London (Figure 5.14). All of these and many more techniques for observation still fundamentally rely on some form of geographic visualization, most often as a graphic or a map. Once GIS modelers can observe the patterns in their own domain, they can begin the process of quantifying what they observe. Next, we examine some common tools for identifying the patterns and for suggesting possible linkages between pattern and process.

Figure 5.14 A portion of Dr. Snow's map of part of London showing the spatial distribution of cholera near the contaminated well.

TOOLS FOR IDENTIFYING PATTERNS

The Landscape

Reading the landscape has a long tradition among geographers, travelers, and field scientists. In many cases, there is no substitute for firsthand experience. Our ability to identify spatial patterns is based largely on our experiences in observing them. The more we observe our environment, the more patterns we begin to see. Domain professionals who are in touch with their environment will most often see patterns that generic GIS modelers who are new to the domain subject matter will miss. As we saw in the first chapter, this suggests that there should be a high degree of interaction between the domain expert and the GIS modeler (assuming they are different people). Both should observe the environment that they are attempting to model whenever possible.

Beyond just visiting study areas, it is important to consider using a wide variety of tools and perspectives to enhance the ability to view them. For example, rather than relying solely on daytime visits, making visits at different times of day or even at night might very well produce different insights than those produced through single-time frame visits. Traffic patterns, criminal activity, plant and animal activity, and many other patterns change throughout the day. Indeed, seasonal visits may also demonstrate differences in patterns resulting from the changes in season. Patterns should also be observed both with the unaided eye and with the aid of binoculars, polarizing lenses, night-vision glasses, and other forms of optical enhancements so that patterns not normally visible with the naked eye can be seen. A difference in perspective may also assist in the discovery of patterns. An environment, whether a physical one or an anthropogenic one, appears quite different from a moving vehicle than on foot. It also looks different from low viewing locations such as in a valley than it does from hilly or mountainous viewpoints. And perspective can also be changed itself to get a different view of the landscape from, for example, a horizontal perspective or a vertical perspective. In fact, the idea of a fly-through, a technique now readily available within GIS, remote sensing, and graphics software packages, provides us an opportunity to vary distance, perspective, and viewing angle very rapidly. This is more difficult in the real world, but varying these aspects will all provide us with additional information about the patterns that are available for analysis inside a GIS.

Literature Survey

An important tool for modelers is the literature survey, which allows them to use the expertise of other domain specialists, either in the absence of other tools for identifying patterns or, more commonly, in addition to them. Surveys of domain literature are normally directed as specifically toward the identified problem, features, or organisms (including people) as is possible. However, knowledge of spatial patterning in other knowledge domains and of the interactions of those patterns and their processes may yield some surprising analogies that can be applied to your future modeling. Take, for example, the landscape ecological literature, in which the pattern of the landscape either reflects historical patterns or results from differences in more recent patterns of plants and/or animals. These scientists speak of "edge species" as species who either live in or forage in sharp landscape edges formed, for example, along the interface between forest and field (Forman 1995). Such patterns, it turns out, share some striking similarities to the activities of some types of criminals, such as career thieves, who search out edges between poor and rich neighborhoods. Such "edge criminals," then, demonstrate a pattern of activity whose predictability can be aided by identifying available edges that tend to encourage it.

There is an ever-increasing body of GIS applications literature within a wide array of knowledge domains, including agriculture, urban and regional planning, crime analysis, health care planning, and ecology. Where once the literature on GIS applications was very sparse, it is now quite robust. This robustness is also linked, unfortunately, to a dispersion of the literature. With the advent of readily available digital search engines, this dispersion is less problematic than it once was. Virtually all of the major literature search engines can now be searched for GIS applications and modeling and produce ample results. Additionally, new collections of digital GIS literature are available through such organizations as the National Center for Geographic Information and Analysis and ESRI's Virtual Campus Library. These collections contain thousands of models based on observed spatial distributions. Perusal of these collections and those found by other digital literature search engines not only will provide an idea of the patterns that exist but also will give ready examples of how they have been modeled in the past.

Knowledge Engineering (Repertory Grid)

Despite the value of firsthand evaluation of areas and of a perusal of the literature, there are often people available whose experiences and sensitivity to patterns within their own specific knowledge domain far exceeds that of the GIS modeler or the inexperienced domain specialist. This is as much true for spatial patterns as for other types of domain knowledge. Unfortunately, this knowledge is often contained within a set of ill-defined heuristics or seat-of-the-pants experience. Obtaining this knowledge is frequently necessary for real-world GIS applications, but unstructured interviews may not be sufficient to obtain it, esecially if the application is time critical. Methods such as the Delphi technique have been applied within groups of individuals to promote discussion and to elicit these heuristics (DeMers 1989). Such approaches work when the components of the knowledge are fairly well defined. When they are not, other, more focused techniques are required.

Within the field of expert systems and artificial intelligence, practitioners have developed a variety of structured knowledge elicitation techniques called repertory grid. Based on Kelly's (1955) work in clinical psychology, these techniques have been employed to identify patterns for GIS implementation (Coulson et al. 1987), especially complex patterns that are not readily visible on the landscape. This method begins by identifying a set of entities or objects, then asks the user to define some constructs (essentially attributes) that characterize those objects. Although the general technique is fairly simple, it has been found to be useful in creating simple expert GIS systems. It seems reasonable that this approach could also be readily applied to the identification and characterization of spatial patterns as well. Additionally, such knowledge engineering approaches are useful for providing an explicit formalization of many otherwise vague, implicit spatial components to domain knowledge. More advanced methods have been applied to acquire more subjective characterizations of regions (Robinson 1990) but are not readily available and are still more theoretical. Repertory grid software is readily available even on the World Wide Web (see http://www.csd.abdn.ac.uk/~swhite/repgrid/repgrid.html).

Maps

Although the repertory grid technique is fairly exotic at least to the GIS community, the map is both an obvious and a readily available source of spatial pattern recognition. If you are reading this book, you are probably aware of the vast array of maps that could illustrate the patterns of a wide array of landscapes and settings. It is not necessary to belabor this, but you should become aware of the many themes, scales, projections, perspectives, dimensionalities, and symbolism types available for viewing spatial settings. Of course, maps can be either analog or digital. In many cases, digital GIS databases contain many thematic maps from which patterns can be recognized. Some of these patterns will be single-theme patterns, whereas others will be visible only when overlaid by one or more additional themes. One common approach to visualization employing overlay techniques uses a perspective view (for example, of topography) with another thematic map (such as land use or vegetation) (Figure 5.15). Many remote sensing and GIS software programs provide methods for perspective views, and some more sophisticated software packages even include for flythroughs that allow for changes in viewing distance, elevation, vertical angle, and azimuth. Such visualization techniques may show the impact of elevation on either vegetation or land use and can suggest some nice predictive model types. This also illustrates the value of a data-rich GIS environment in which to perform simple spatial data exploratory analysis.

Aerial Photography

It is no secret that many maps were originally derived from interpretations of aerial photography. Access to the original photography itself, however, allows users to decide which objects to concentrate on and which patterns might emerge, rather than rely on the cartographer to perform these important tasks for them. Aerial photography interpretation skills can be quite useful for identifying visual patterns.

Figure 5.15 Perspective view of a mountain on which the vegetation is overlaid. This illustrates the idea of vegetational zonation.

Through employment of evaluation of tone, color, texture, setting, scale, association, time, and other interpretation factors, not only is discovery of patterns made relatively straightforward but also discovery of a context within which they exist is made apparent. Such a context is vital to making functional links among pattern elements and among thematic layers.

Satellite Imagery

For area coverage that is generally larger than that available or economically viable through aerial photography, satellite imagery may prove a useful alternative. Satellite remote sensing has developed both technologically and intellectually; with a wide range of spectral, spatial, temporal, and radiometric resolutions, and area coverage; there is ample opportunity to include some forms of satellite remotely sensed data in the observation of patterns, especially for a synoptic view of large areas (Jensen 2000). For land and ecologically related GIS models, the utility of satellite data cannot be underestimated. Patterns that do not show up at large scales (say, 1:24,000 aerial photography) may be readily visible on 1:1,000,000–scale *LANDSAT* imagery.

Beyond their synoptic view and lower cost per unit area, remotely sensed data, being available in digital formats, allow appropriate digital-enhancement software to extract patterns that are not readily visible. For example, high-pass filters enhance edges; low-pass filters reduce noise, making other general patterns easier to see; and directional filters stretch contrast. A host of other image-enhancement techniques are specifically designed for teasing patterns from satellite remotely sensed data. One additional important characteristic of satellite remotely sensed data is their frequency of availability (temporal resolution). The more often these data are available, the more often they can be added to a spatiotemporal database. In this way, patterns of change may emerge that would otherwise remain hidden.

Statistical Techniques

But of all the tools available for recognizing patterns, especially patterns that are not visible on the landscape, statistical techniques are among the most powerful—and among the most often ignored by GIS professionals. Of particular utility is the vast array of correlation and regression techniques that can be employed to determine relationships among variables. Even simple descriptive statistics like spatial mean, median, and mode can show patterns by allowing the user to identify summaries of spatial data. These are more useful for GIS and spatial pattern recognition when weighted by area.

Beyond measures of central tendency, measures of dispersion (including spatial dispersion measures such as nearest neighbor and weighted mean center) and inferential statistical techniques can be employed on a sample of your spatial data set to determine whether they might be indicative of the patterns you might find in the larger data sets. These approaches are excellent means for rapid prototyping and characterization of the patterns of your larger study areas and their data sets. Becoming proficient in statistical analysis improves your ability to identify individual and group patterns. More importantly, statistical testing is an excellent method of predetermining the interactions of spatial factors that will be employed in your stochastic GIS models.

RECOGNIZING THE SPATIAL INTERACTIONS OF THE PROBLEM: FROM PATTERN TO PROCESS

Among the most important tasks that we will need to perform prior to modeling is hypothesizing the possible processes that relate to the distributional patterns we observed. In some cases we will be trying to determine the underlying processes that created the patterns, whereas in others we will be trying to evaluate the effects of existing patterns on ongoing processes. To evaluate either requires that we go beyond an ability to recognize that patterns exist and begin to describe what the patterns are. We will need to describe the geometry of individual objects and the spatial arrangements of groups of objects. Measures of absolute and relative size, orientation, circuitry, connectivity, isolation, dispersion, density, shape, spatial integrity, and many more qualities provide us with quantifiable descriptors of the patterns we recognize. The evaluation can be an examination of the geometry and arrangements within a single thematic map, or it can be a comparison of entities between two or among multiple thematic maps. Point, line, area, and surface objects can all interact with one other, and these interactions are often what we are attempting to model with the GIS. At this point, it is prudent to review the functional capabilities of GIS for such tasks by surveying some of the introductory texts on the subject (e.g., Chrisman 1997, Clarke 1999, DeMers 2000a, Heywood et al. 1998).

As you might imagine, there are millions of wildly different objects and even more millions of possible combinations and permutations of geometries and arrangements of these objects. So how can we expect to know or even surmise all of the possible causes for and impacts of so many instances? With the question posed that way, the simple answer is: We cannot. But we can make some generalizations about the limited geometries and arrangements we can describe. This is at the heart of the geography behind GIS—the ability to observe, describe, and evaluate the interactions of earth's objects, features, and occupants and their geographic space. Examples may give us a chance to exercise our spatial thinking, especially with regard to the relationships between spatial patterns and the causal or resulting processes. Consider Table 5.1, which lists just a few of the many possibilities.

Notice how each object carries with it a dimensionality (point, line, area, or surface), a character being examined (e.g., volume, slope, aspect, size, orientation), a

TABLE 5.1 Some Objects that Demonstrate Identifiable Spatial Dimensions, Measurable Characteristics, and Cause-and-Effect Relationships

Spatial Dimension	Object	Character	Measure	Cause	Effect
Surface	Slump block	Volume	Morphometry	Gravity/fluid input/pressure	Slope instability
Surface	Topographic ridge	Slope/aspect	Angular degress/ azimuth	Uplift	North versus south vegetation
Area	Forest patch	Size	Perimeter or per- mimeter/area	Forest clearing	Interaction with matrix
Area	Linear forest patch	Orientation	Long axis azimuth	River corridor	Migratory bird roosting
Line	Hedgerow	Extent	Length	Human	Animal movement
Line	Road network	Connectivity	Alpha Index	Human	Traffic flow
Point	Gopher holes	Density	Number/unit area	Colonization	Competition
Point	Fruit trees	Arrangement	Nearest neighbor	Planting	Crop efficiency

specific measure associated with that character, a potential cause, and possible effect. The cause can be thought of as an hypothesis of the reasons behind the development of the patterns. As a hypothesis, it should be testable at least against some random process (effectively, the null hypothesis). Here is where statistical testing can be very effective in identifying the functional relationships between existing patterns and past processes. A similar approach could be taken in examining the effects of existing patterns on ongoing processes. In each case, the pattern is tested against some observable resulting process. For example, if the overall length of a hedgerow increases, a reasonable hypothesis is that there will be a measurable, predictable increase in the numbers of species that favor edges (edge species) (DeMers et al. 1996). The null hypothesis would be that such an increase in edge species will not result. Again, this shows the utility of applying statistical testing to provide a working knowledge of functional relationships between pattern and process prior to creating a GIS model.

Of course, this does not negate the possibility of using the GIS itself to test these hypotheses. In fact, employing statistical testing within a GIS using a small subset of empirical thematic data is an effective way of using GIS as an inferential statistical testing tool. Once spatial relationships are established for a sample of the data and the confidence limits are known, you can then implement the model for the entire database, on the basis of the established relationships. This also provides you with some measure of the confidence limits for your GIS model itself. A now classic example of using raster GIS to implement a predictive statistical model is the timber breakage model used by Tomlin (1981). By employing a regression equation from within the Map Analysis Package, Tomlin extended the typical regression equation to a spatial domain, thus removing the necessity of performing the regression test prior to GIS implementation.

Although this brief discussion of statistical analysis does not enumerate all the possible ways in which such tests can be performed, it does point out the importance of establishing the functional relationships among patterns within or prior to GIS modeling. Additional tools such as logit modeling, sensitivity analysis, and autocorrelation can be used to identify and quantify functional relationships (e.g., Algarni 1996, Clark et al. 1993, Johnston 1992, Lowell 1991, Pereira and Itami 1991). The advent of a burgeoning set of spatial descriptors, especially those found in the landscape ecology, health care mapping, and spatial crime analysis literature, has prompted an equally growing need to identify and quantify the causes and results of these patterns. The quantitative measures of pattern are a necessary first step, but without linking them to causation, we are unable to build effective, real-world GIS models, whether they are designed to describe a situation or to predict new ones.

TYPES OF GIS MODELS

Introduction

A classification of GIS models, like a classification of anything, is based on a preselected set of criteria. There are many ways of classifying GIS models—so many that it can get very confusing. In fact, Berry (1987, 1997) has separated spatial models from cartographic models, whereas many authors do not. My purpose here is not to add confusion to your already complex modeling tasks by creating yet another set of classifications but rather to examine some basic terminology that is being used among GIS modelers so you can communicate effectively with them. Additionally, the classifications will provide a structure for the modeling tasks by describing substantially different ways of modeling based on selective purpose, varying methodologies, and often fundamentally different logics. It is important to note that although some of these classifications are unique, many cross over and intermix, resulting in a relative inability to create a classification hierarchy that was originally attempted by Tomlin (1990). This confusion can best be assuaged by determining the specific utility of each classification scheme and treating each in turn, as was suggested by Berry (1995). The following paragraphs use this approach, providing important considerations for modeling within each model class. As you read them, you should spend less time on the classification itself and more on the modeling tasks and attendant thought processes specific to each. I limit the classification of GIS models to three fundamental approaches: (1) purpose, (2) methodologies or techniques, and (3) logics. Again, remember that none of these classifications is entirely independent of any of the others.

Models Based on Purpose

If there were a single well-accepted hierarchy to GIS modeling tasks, it would most likely begin with a focus on the overall purpose for which the model is to be developed. As with most other classification methods, one based on purpose is neither discrete nor binary but rather demonstrates extremes in a continuum of possibilities. On the one extreme, we have models whose sole purpose is to describe. These are called *descriptive* models. At the other extreme are GIS models whose primary purpose is to prescribe best uses of existing land resources on the basis of evaluation of known or predicted circumstances. These are *prescriptive* models. These two seemingly discrete model types, although no less a continuum than others, seem to have become accepted as the two most basic types of GIS models within the literature. We begin with the one that is most often the more basic of the two—descriptive—and increase in complexity to prescriptive models.

Descriptive Models As the term implies, descriptive models are *passive,* primarily designed to provide a description of parts or all of a study area under examination. The description can be simple or complex, single or multitheme, preparatory to final modeling or a solution in itself. It is sometimes difficult to separate the term *descriptive GIS model* from other model types because the map itself often describes, in explicitly spatial terms, conditions as they exist, could exist, or should exist. The terminology depends more heavily on the purpose for which the final outcome map, graphic, or other output is to be used than on the output itself. Descriptive models, then, describe conditions as they exist. In other words, they most often answer the "what is" question, rather than the "what should be" question. In some cases, the descriptive model may describe conditions that might also fit selective uses of the land. In such cases, the descriptive model might be answering the "what could be" question by simply describing the conditional fit rather than suggesting actual use.

In their simplest forms, descriptive models attempt to quantify an existing map or set of maps on the basis of any of the functional operations we examined in Chapter 4. In the first case, this type of model attempts to describe the geometry of the components map or maps. This geometry can range from simple measures of length, width, perimeter, area, circuitry, and many more, to more complex, integrative measures such as perimeter-to-area ratios, nearest-neighbor metrics, isolation, and other more topological measures inherent in the cartographic document. Many of these were described earlier in this chapter. The importance of describing or quantifying the geometry of a map is that it allows us an opportunity to isolate patterns within the overall map; to discover patterns that may not, without quantification, be visible;

or to compare patterns from one map to another or from one part of the map to another. Because each pattern is a result of one or more underlying processes, the description of pattern also helps us gain insights into these processes. Dispersion patterns, for example, are often used to relate pattern with process. Clustered dispersion patterns are often a result of nonrandom processes, as are uniform dispersion patterns as one might find in an orchard or row crops, whereas random dispersion patterns are most often a result of stochastic or statistically random processes. These patterns and their associated processes may also be related to other patterns on other themes. And the processes that formed the additional patterns may also be related to the original patterns.

Among the more powerful capabilities of the descriptive GIS model is its ability to go beyond descriptions of geometry to integrate or synthesize often seemingly disparate spatial data. In this context, the descriptive model could also be called a *synthetic* GIS model not because it attempts to describe a situation by examining a single map element or even a single map but because it often merges multiple themes to evaluate possible spatial relationships. In the synthetic approach, successive themes are combined, one at a time, to determine the degree of spatial association each might have in an overall description of existing conditions. Descriptive models are a mainstay within the scientific community. Scientists' training encourages them to follow a pattern of behavior, called the scientific method, that begins by first making observations of patterns, the cause of which will later be hypothesized about, will be rigorously tested, and may evolve into theory or, if proven immutable, become scientific law. Although the GIS is not, in its current form, a particularly good tool for testing hypotheses, it is a very adaptable tool for creating testable spatial hypotheses (Aspinall 1994).

An alternative to the synthetic type of descriptive model is the *deconstructive* type. For determining the sensitivity of certain factors in a descriptive model, it might prove useful to remove each one at a time, examining the final result as each part is removed. This is similar to a stepwise backward approach to regression modeling as opposed to stepwise forward. In stepwise backward regression, one attempts to remove individual independent variables to ascertain the impact each has on the final regression coefficient of the model. Although descriptive GIS models do not currently have a coefficient as a final result, it is still possible to at least see whether particular spatial variables correspond spatially by using a simple test of spatial correspondence (Muerhcke and Muerhcke 1999). And, of course, as with statistical correlation and regression analysis, the mere existence of spatial correspondence does not prove causation. It shows simply that the selected variables occupy some portion of the same geographic space. Such spatial correspondences can, however, be highly suggestive of such causation if carefully chosen (Sauer 1925).

Prescriptive Models   At the other extreme of our classification continuum is the more *active* prescriptive model. In its purest form, the prescriptive model is designed to impose a best solution for problems in which a description of existing conditions is insufficient as a decision aid (Tomlin 1991). Just as a physician would first describe symptoms of a disease or other medical condition and then, on categorizing these symptoms (assigning a name, usually), would take of the next step of prescribing the best medicine or treatment to cure the problem, our next step would be to "prescribe" the best solution to geographic problems. In GIS, such scenarios might more aptly be applied to answer such questions as: (1) What is the best location in which to site a factory? (2) Where is the most likely location for finding a serial killer? (3) What is the most likely place to reintroduce aplomado falcons in the southwestern United States? In short, the prescriptive model is more closely associated with answering the "what should be" type of question.

As with any prescription, there isn't always a perfect solution for a given question. In this case, there are generally two approaches. One is to select a best solution on the basis of the best available data and the constaints that currently exist or are expected to exist (if the model is to be *predictive*). This first type is most often used when the individual constraints driving the model can be or are limited to Boolean conditions (i.e., goodsoils versus badsoils or goodzoning versus badzoning). These models are actually pretty rare and are most often implemented in the absence of specific factors with a range of conditions. The second approach is similar, except that it provides a range of possible solutions, some better fitting certain criteria than others. This approach is best applied when more information is known about the conditions of each included factor. Because there is a range of possible factor ratings and weightings, there is a wider range of factor sensitivity and thus a greater opportunity for effective, if not optimal, solutions. In this way, if some unforseen economic or political power should preclude the use of the best site, others would be available for use.

Among the most important characteristics of prescriptive models is their ability to derive a solution, not just to describe what is already there. As such, the prescriptive model tends to be more adept at prediction. Stated differently, if you have a *predictive* GIS model, it is more likely to be a prescriptive one than a descriptive one. But this does not preclude descriptive GIS models from having at least some predictive qualities. It is also important to understand that not all prescriptive models are predictive. Generally, for a model to be effectively predictive (prescriptive), it is very important that the processes that link the themes are explicitly and very completely understood. Such models also typically contain some dynamic elements and may require special database structures (e.g., the cellular automata) or even special computer processors for more complex types (Costanza and Maxwell 1991). Perhaps the classic examples of *dynamic* predictive maps are those that include dispersion or movements of ideas, creatures, or processes. Fire modeling is among the best known and most obvious of such predictive models.

In his original design, Tomlin separated prescriptive models into two types: *holistic* versus *atomistic*. Holistic models are those that evaluate a scenario in its entirety. They require a complete, overall understanding of both the processes and the thematic contents of the maps. Such models are rare, partly because there are few situations in which the overall complexities of systems are fully understood, and partly because they are very difficult to verify and validate. The more common type of prescriptive model is the atomistic type that breaks its processes and its themes into categorical and functional groups. By its very nature, it is the kind of prescriptive GIS model that readily lends itself to compartmentalization. It proceeds step by step, isolating individual elements as it proceeds. Because of that, it is far easier to conceptualize, to formulate, to flowchart, to implement, and to verify and validate.

In turn, atomistic prescriptive GIS models can be broken down into two additional categories: *heuristic* and *algorithmic*. Heuristic model types require either called-on experience or seat-of-the-pants experience. These types of experiential knowledge are often poorly documented, rarely formalized, and very difficult to obtain. Earlier, we talked about the use of knowledge acquisition strateties for obtaining GIS modeling information. This is the classic type of GIS model requiring this knowledge, necessitating unique knowledge acquisition strategies. Actually, on acquiring these heuristics, one most often needs to formalize them into a more explicit recipe for decision making. In other words, heuristic models are actually easier to solve if they are transformed in to *algorithmic* models.

Algorithmic prescriptive models most often take the form of a set of rules that explicitly relate elements of each thematic map. They relate these elements and these maps in a specific order or sequence, often in a hierarchical fashion, that both is representative of the real-world processes and allows for reversal of the process for model verification. In fact, these properties nearly define what a GIS model is—an ordered set of map operations designed to represent real-world settings.

Models Based on Methodology

As with virtually all other model types, the methodology of cartographic models is either *stochastic* (based on statistical probabilities) or *deterministic* (based on known functional linkages and interactions). Stochastic models are linked to the statistical criteria most often used in non-GIS models. For example, models based on statistical measures of central tendency are of necessity driven by the central limit theorem. An extension of this theorem for predictive modeling is regression analysis. One classic example of the use of regression modeling is Tomlin's (1981) predictive GIS model of timber breakage during harvest. The model employs a regression equation on a cell-by-cell basis. This is effectively a spatial regression model. Another example of effective stochastic modeling is the use of logistic regression (categorical regression) to predict the presence or absence of creatures in an underlying environment. Models like this include those examining squirrels (Pereira and Itami 1991), bears (Agee 1989, Clark et al. 1993), desert longhorn sheep (Dunn 1996), deer (Chang et al. 1995), and birds (Miller et al. 1989), each of which exemplifies how statistical techniques can be linked with other GIS functionalities.

Although stochastic GIS models assume things are distributed on the basis of statistical likelihood, deterministic models assume direct functional linkages. Models such as those involving hydrological flow prediction (Chase 1991), pollution evaluation (Gros et al. 1988, Haddock and Jankowski 1993), and soil loss modeling using the universal soil loss equation (Battad 1993) are good examples of how a knowledge of the environment can be modeled with deterministic methods. A primary consideration for these and all deterministic models is that a well-defined cause-and-effect relationship exists and can be identified.

Models Based on Logic

We have examined how GIS models can be based on purpose as well as on the methodology applied to their creation, but the method of logic applied in the conceptualization and formulation of the model is equally fundamental. There are two primary forms of logic that are traditionally employed: *inductive* and *deductive*. The inductive method attempts to build general models based on individual data or instances. For example, by collecting or obtaining data and information on mountain lion habitat use at a number of individual locations, and by summarizing those localized conditions, one can begin to create a general model of mountain lion habitat use in a region. In short, the inductive approach moves from specific elements or instances to a general model, usually employs many *empirical* tests to gauge the viability of each factor, and typically uses a trial and error approach.

This approach is often useful if we are unaware of the general conditions or rules under which our subject or subjects operate. In some circumstances, especially within a data-rich environment, many previously unknown and important factor interactions can be identified by such a *data mining* aproach. To some, the inductive method can be a bit disconcerting because it seems to eliminate the hypothesis testing we are so familiar with in the scientific method. This not true, however, because each of the factors employed is tested. Its advantage over the deductive approach is that it is far less common that we know how all the processes function than is practical for many modeling environments.

But models that build on deductive logic are straightforward, easier to understand, and far more purely algorithmic (atomistic) than inductive models. Deductive logic moves from general to specific. In GIS modeling, what this implies is that we have a substantial preliminary knowledge of what factors are important, how they interact, and which are most important before we conceptualize, formulate, flow-chart, and even implement the model. The best candidates for such modeling are those that already have a somewhat formalized set of criteria, weights assigned to each, thematic map data that represent them, and standards for how they are combined, algorithmic models such as the simple additive LESA (Land Evaluation and Site Assessment) model (Williams 1985), the statistically based timber breakage harvest model (Tomlin 1981), and models employing the universal soil loss equation (Battad 1993). GIS models that employ existing aspatial mathematical or statistical models and add the spatial dimension through the use of a raster-based GIS are some of the more obvious examples. Some of these can become quite complex. Some even require the implementation of parallel computer processors to complete their calculations (Costanza and Maxwell 1991), yet they still remain more easily understood than many less algorithmic inductive model types. In fact, deductive models—even very complex deductive models—are easier to explain to clients and are therefore much easier to verify and to validate than are inductive GIS models.

Chapter Review

An essential skill for geographic information system (GIS) modeling is the ability to recognize, identify, and interpret geographic patterns. In essence, it requires spatial thinking. The GIS model may be used to evaluate, describe, or combine either visible patterns (those that are fairly obvious) or functional patterns (those that may be visible only through special means of sampling or unique tools of observation). The primary tools for identifying patterns include visits to the landscape, examination of other research through literature review, interviews and other knowledge engineering techniques, examination of maps, aerial photography, satellite imagery, and descriptive and predictive statistical techniques for relating dependent and independent spatial variables. Statistical techniques, frequently combined with GIS analysis itself, can often link the quantified geometry of individual or multiple thematic maps with functional processes that might be responsible for them. These hypotheses can be an end in themselves or they may be used as the starting point for a GIS model.

There are many kinds of GIS models, depending on how they are classified. On a simple level, the model types are divided into three classification schemes. The first is based on purpose and includes descriptive models and prescriptive models. Although these two types of models are actually part of a continuum, descriptive models are primarily designed to answer the "what is" question, whereas prescriptive GIS models most often answer the "what should be" type of question. The second GIS model classification scheme is based on methodology that includes stochastic (statistical) versus deterministic models, where cause-and-effect relationships can be effectively identified. The final class of GIS model types is based on the type of logic used to implement the model. In this case, some models are inductive, in that they try to make generalizations based on subsets or samples of the population of the data, or deductive, where knowledge of the overall situation can be used to predict individual conditions.

Discussion Topics

1. The U.S. Forest Service is building a generalizable predictive geographic information system (GIS) model of forest fire potential based on factors such as forest use,

fuel types and buildup, locations of campgrounds, surveys of visitors, and many others. Aditionally, the model is also designed to predict the spread of fire on the basis of wind speeds and directions, topography, humidity, and many other factors. Discuss the types of GIS models we have just examined and categorize this model in general and its component parts in particular.

2. A new acquaintance does not understand what the problem is with creating GIS models. After all, it is just like using any other piece of software. Once you learn which buttons to push, the rest is just rote. Provide a modeling scenario in which your new acquaintance will be forced to think in explicitly spatial terms. The purpose of the exercise is to make you think spatially to develop the scenario and for your friend to do so to discuss its solution.

3. Discuss the role of repertory grid methodologies in acquiring knowledge about the relationships between thematic map representations of spatial data and the functional operators. What other knowledge acquisition techniques are available for this same task? What are the advantages and disadvantages of each?

4. Consider the following adjectives that can be applied easily to different types of GIS models and discuss how they would fit within the three basic classifications employed in this chapter:

   a. Predictive

   b. Simulation

   c. Spatiotemporal (dynamic)

   d. Land capability

   e. Land suitability

5. For at least three different knowledge domains, provide some appropriate visualization techniques, statistical approaches, and literature sources to provide the basis for spatial understanding. Example knowledge domains might include criminal justice, defense, land planning, atmospheric modeling, habitat evaluation, site selection, health care provision, real estate, and insurance.

Learning Activities

1. In your nearby community, go for an undirected road trip, carrying with you a camera (even one of the disposable ones will work), with at least 24 frames of print film available. Your mission is to photograph your environment as you observe it firsthand, in particular to observe and document patterns. After developing your film, provide a 3″ × 5″ index card for each photograph describing the nature of the environment you have photographed and the patterns you observe. Create a poster of your photographs and index cards that you will share with your other classmates.

2. From your photographs and those of your classmates, create a table similar to Table 5.1 that shows the spatial dimension, object, character, measure, possible cause, and effect for each of the objects you identified.

3. Collect 10 to 20 GIS modeling articles at random from a range of journals in the literature (especially the professional journals) without selecting them by type (descriptive versus prescriptive), by methodology, or by logic. Create a table with five columns. The first column should be an abbreviation of the article title, the

second column should be the name of the journal, and the three remaining columns should be model categories (purpose, methodology, logic). Assign each article to the first and second columns. For the model types columns, classify each article and place that information in the appropriate column. From this, suggest which journals seem to specialize in particular GIS model types. Compare your results with those of other students in the class. Does this suggest where to obtain examples of particular types of GIS models?

## CHAPTER 6

Conceptualizing the Model

On completing this chapter and combining its contents with outside readings, research, and hands-on experiences, the student should be able to do the following:

1. Define the goals of GIS models given specific scenarios and model types

2. Using existing published GIS models, deconstruct them to reveal the conceptual design elements from which they were created

3. Specify the precise spatial information products that should result from selected GIS modeling goals

4. Break the spatial information products into their component parts, thus creating compartments from which submodels can be developed

5. Explicitly define the relationships and connections among individual component parts of a GIS model

6. Define cartographic representations and thematic maps necessary to represent individual types of model components

7. Evaluate which model factors are aspatial and are unavailable for input into individual GIS models

8. Define when aspatial data can be included as operators among cartographic representations of spatial data

9. Enumerate ways of accounting for missing thematic map data

10. Enumerate ways of defining and using spatial surrogates for either missing or aspatial thematic data in GIS models

11. Define explicit sources of data, data types, and data input methods that will be used for formulating your GIS model

INTRODUCTION

In Chapter 5 we examined a number of ways of thinking of GIS modeling and general types of models that a GIS is capable of creating. In particular, we looked at how models range from purely descriptive to the often more complex prescriptive models and combinations of these types of models. These structures are essential in conceptualizing and formulating models because they give us a framework for deciding what the final outcome of our modeling process will be and how we intend to arrive at that outcome. But even this basic framework is only the first step in actually creating a GIS model. Once the final outcome is decided on and the methodological approach is established, we need next to determine specific strategies for implementing the model. This is a somewhat more specific examination of the basic elements that compose the model you are building. The following pages will assist you in selecting appropriate logics and translating them to a compatible form of expression. We will look at the utility and potential pitfalls of each and begin the process of map element identification. Later on, in Chapter 7, we will formalize this through the use of model flowcharts. For the time being, however, we will stick to the more nebulous but no less important tasks of conceptualization and formulation.

Discussing the ideas of GIS model conceptualization and formulation is nearly impossible in the absence of concrete examples. Tomlin (1990) recognized this in the creation of his much-cited text on GIS modeling by using a single database throughout the text. His idea was to allow the user to have a common set of data and a limited set of themes within which to develop his modeling language. Although this approach was practical for illustrating how map algebra works, I think that the increased numbers and variety of models, modelers, topical domains, contexts and increased levels of model sophistication and audience sensitivity all beg the need for a somewhat more realistic appraisal of GIS modeling using a real GIS data set and a real model, supplemented by examples whenever possible. What I hope to do is expand on the ideas and concepts first voiced by Tomlin.

Whether your model is to be a purely descriptive model or a prescriptive one, it is necessary to conceptualize the model first so that you can effectively formulate it into its component parts. Model conceptualization is a more general form of the actual formulation process. As with its cartographic equivalent, the purpose of conceptualization is to envision, often by comparing your intended model with examples from the literature, how your model is supposed to work. Once a general conceptual vision of the model is established, we can move on to the actual formulation and flowcharting of the model that we will cover in the next chapter. We begin by examining our goals, then go on to breaking down our problem into compartments, systematizing the compartments, identifying their spatial dimensions, and finally, identifying possible data sources and types (Figure 6.1).

DEFINING YOUR GOALS

Proper design, whether it is related to creating a bridge, building a house, or putting together a GIS model most effectively, works backward, beginning with the desired end product and ending with an explicit definition of the necessary components and interactions. In GIS design, the goal is most often called the spatial information product (SIP), as we have already seen (Marble 1995). This acknowledges that the final outcome nearly always takes the form of information, demonstrating both a spatial context and some model-specific subject domain context, resulting from some integration and/or manipulation of spatially explicit thematic data. In some cases, a single SIP is all that is needed. For example, you may be trying to determine the best location for siting a sanitary landfill on the basis of soils, surface and subsurface hydrology, land availability, zoning, and accessibility considerations (a prescriptive model). Alternatively, you may be creating a more flexible model whose SIPs are less well defined (as in creating a descriptive GIS model for data mining), requiring mul-

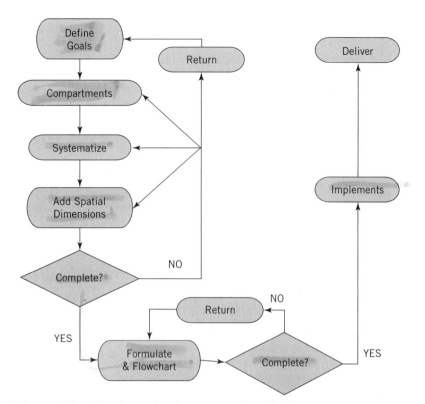

Figure 6.1   Flowchart showing a general modeling process.

tiple output maps or even multiple output types. Or you might be designing a model that has more than one audience (e.g., one in which land use decisions affecting divergent groups may be required to enable spatial conflict resolution). No matter which of these defined scenarios you employ, or which of the many alternatives of your choosing, it is always best to decide on the final outcome before proceeding. At the same time the modeler should be careful not to prescribe the outcome to fit expectations, but rather to provide an objective result based on the data.

Although descriptive and prescriptive models operate differently and although the goals and expected SIPs are often different, nearly all prescriptive models have as their foundation a substantial descriptive component. For this chapter, we will focus on the descriptive model for this reason, and we will see in the next chapter how the models start to bifurcate when we begin to formulate and flowchart them. As we saw earlier, descriptive cartographic models tend to be more synthetic than analytical, and many prescriptive models begin with some form of synthesis and finish with more analysis. In other words, descriptive models tend to put data together within a domain context that allows the user to make decisions from them. GIS models of either general type might embody well-known, even obvious, data types that instantly come to mind. This suggests that by simply collecting all the existing data for a particular study area, we merely need to find appropriate means of connecting them to put them into context. Such an approach, although seemingly efficient, often leads to omission of important data, imprecise or inexact thinking, and indefensible models based on natural or cultural bias. In fact, such models often reflect a preordained outcome rather than a well-thought-out strategy.

With the proliferation of geocoded data sets at varying scales, it would be easy to begin by anticipating how these thematic maps might be applied to the model. Although this might seem easy and even reasonable, it is very important that you resist this temptation unless your primary purpose in modeling is for data mining and GV for hypothesis formulation in a data-rich environment. Under most real-world GIS modeling settings, this is not the case.

I do not want to belabor this point, but I think most texts do not emphasize this latter point enough. Among early GIS applications researchers, a common concern is that the problem they are attempting to solve or the study area they have selected for a particular model may lack the data sets necessary for completion.

Some major reasons for ignoring existing data sets as a starting point for GIS modeling are that

1. Many data sets not specifically compiled for your model will not have the necessary data integrity, accuracy, scale, classification systems, and so forth

2. Many data sets that are not model specific will contain too many themes, often suggesting factors that are irrelevent to the model and increasing data storage and cataloging requirements

3. Alternatively, many data sets are incomplete for specific models

4. Data sets can often bias your thinking from both a methodological and from a conceptual level

5. Area coverage and sampling procedures are often inadequate for specific modeling tasks

One simple alternative to collecting available data sets for your area and trying to decide what to do with them is to formally state what it is you intend the model to do before you collect any data at all. If, as is often the case with descriptive models, the purpose is to synthesize criteria for such tasks as describing and ranking the potential capability of a land resource base for tasks like powerline corridors, solid waste facilities, or housing developments, the criteria ultimately drive the model and suggest the appropriate data. You should note here that these models, although descriptive in nature because they answer not only the "what is" question (i.e., What is the current capability of the land?) but also begin to consider the potential "what could be" question, such models begin to move along the continuum toward answering the "what should be" question. Again, this shows how a prescriptive model always contains a descriptive component.

Let's take a look at a now classic example descriptive raster GIS model from the literature—the U.S. Department of Agriculture's (USDA) LESA model. The LESA model was originally an aspatial—or at least site-by-site—evaluation of lands at the county level in the United States. It was designed to provide decision support for planners concerned over the proper allocation of agricultural lands for nonagricultural uses. This model is more closely associated with descriptive tasks than prescriptive in that it does not technically allocate any specific lands but rather rates them numerically on the basis of a set of criteria based on negotiations among the USDA participants and local or regional planners.

Williams (1985) published a relatively detailed description of the model, particularly concerning the site assessment portion that focuses on the nonsoils segment. The land evaluation portion is concerned solely with the quality of the soil for agriculture and is based on a selected indicator crop for the area. Such tasks usually require more non-GIS activities than GIS activities, and the soils map for the county is reclassified on the basis of numerical rankings of each soil type. Evaluating the soils for uses other than agriculture requires far more information about infrastructure, socioeconomic factors, planning and zoning regulations, and many other factors.

Rather than immediately turning to the published work, imagine that you are trying to create a site assessment model to protect against improper use of agricultural lands near the urban areas in your own town or city. You begin, as we have already mentioned, with the goal first, then work backward. What are the essential goals of a site assessment system? Obviously, they are to assess the quality of a site—but quality for what? Because your goal is to protect agricultural lands from conversion to nonagricultural purposes whenever possible, you need to look at what factors might influence such decisions. This requires that you know something of the planning process, about agriculture, about land conversion, and a host of related topics. But these ideas are not necessarily unique to agricultural land conversion.

First of all, in the land-planning scenario, you know some pretty basic things, such as the following:

- **The basic premise of the model—that there is some level of demand for nonagricultural uses of the land resource base.** When demand for nonagricultural uses increases, this puts pressure on the planners to allow more agricultural land to go out of production in favor of nonagricultural uses (Figure 6.2). Such pressure is often a result of increased urban population, improved nonagricultural economies, and higher prices for land uses other than agriculture. Demand for nonagricultural uses of the land is not an explicitly spatial factor and may require you to use surrogates to incorporate it into the model.

- **Someone already owns the land.** Land ownership is an important factor in land conservation because it either enhances or restricts the potential for nonagricultural use. If the land is owned by a farmer who is actively cultivating it, this forces the potential nonagricultural user to apply political or economic pressure to either force or encourage the current owner to sell the property. This would act as a disincentive for land conversion. Alternatively, if the current owner is not cultivating the land, and especially if the owner is in fact planning alternative uses, this would act as an incentive to grant nonagricultural uses. Land ownership links the actual areal parcels to individuals, governments, or coporate owners, making it relatively easy to convert it to GIS themes.

- **The land has a particular size and/or configuration.** The location, juxtaposition, or size of the land parcel in the county has much to do with its viability for either agriculture or nonagricultural uses (Figures 6.3 and 6.4). Large industries, subdivisions, and large-scale agriculture require large parcels of land. The exact size may differ from place to place. Additionally, some parcels of land are fragmented, whereas others are contiguous. Some are available in rectangles or other configurations that possess more interior than edge, whereas others are long, skinny, and even sinuous. Each of these has its own benefits and liabilities for agricultural and nonagricultural uses. This gives you incentive to use your GIS to analyze size and shape of all parcels in the county.

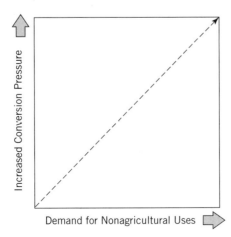

Figure 6.2 **Possible relationship between increased pressure to convert land to nonagricultural uses and the demand for nonagricultural uses.**

Figure 6.3 Possible relationship between the size of a farm parcel to the viability of that parcel for agriculture.

- **The land has a known current use.** You saw in the land ownership factor earlier that the owner may be using the land for either agriculture or nonagricultural activities (Figure 6.5). There is therefore a direct linkage beween the ownership of the land, its current use, and proposed uses of the land. High-energy land uses (those requiring substantial inputs of money, material, and construction) tend to remain in those land use types. As such, there is a disincentive to change these land uses to alternatives unless there is a clear benefit to doing so.

- **Land has some aesthetic value (even as cropland).** Among the more common problems associated with land planning is the NIMBY problem (not in my backyard). Although we will not deal with the many possible conflict resolution problems in this chapter, you must keep in mind that some land uses are, by nature, more aesthetically pleasing to some than are other land uses. This is a difficult factor to quantify and has no explicit spatial dimension.

- **Crops and nonagricultural land uses need water.** Although agriculture requires water, this may be available through precipitation rather than irrigation. Nonagricultural uses of the land may, however, require the use of municipal sources of water for day-to-day operations. Municipalities generally have detailed surveys of their water lines, and parcels targeted for nonagricultural uses that are located near the water lines have an advantage over those that are not. Again, this provides us with a way to rate our parcels on the basis of distance to sources of municipal water.

- **Roads are essential for transportation.** Whether for agriculture or nonagricultural uses, it is essential that roads exist for getting to and from the land parcels (Figure

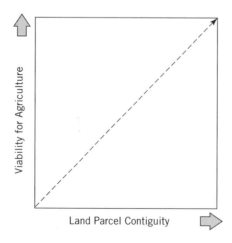

Figure 6.4 Possible relationship between the contiguity of land parcels and the viability of land for agricultural use.

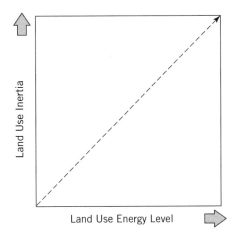

Figure 6.5 Possible relationship between the energy associated with land use type and the inertia of the land to stay in that land use type.

6.6). For agriculture, gravel or dirt roads will suffice for the movement of farm implements; for many nonagricultural uses, the existence of paved roads and, in some cases, multilane highways for access greatly enhances their viability. As such, the existence of such roads near the proposed sites is advantageous for nonagricultural uses and will likely reduce the likelihood of maintaining agriculture in those parcels.

- **Many nonagricultural land uses require electricity.** Although agriculture has need of electricity, such a need is minor compared with the needs of many nonagricultural uses such as business, industry, or residential. As such, the availability of electrical service favors nonagricultural uses over agriculture.

- **Most nonagricultural land uses require sewerage.** Nonagricultural uses require the disposal of liquid waste. This is generally done either by connecting to the available municipal sewage system or by using on-site septic systems. The ability to connect directly to the municipal sewage acts as a positive incentive for the incorporation of nonagricultural land uses.

- **There may or may not be legal restrictions on the land (such as easements, zoning regulations, and the like).** Most municipalities have a set of zoning regulations or easements that restrict the use of the land for particular purposes. There is often a set of planning heuristics behind what restrictions apply where, and some communities are more restrictive than others. Whichever the case, if certain land uses are prohibited on a given site, or within a certain distance of a site that is

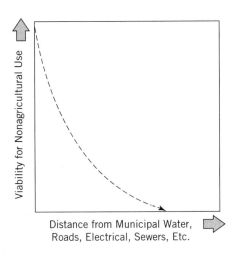

Figure 6.6 Possible relationship between the distance to municipal facilities and the viability for nonagricultural use.

zoned for a particular use, this could prohibit conversion of agricultural land in some areas and favor it in others. These factors are explicitly spatial.

• **There may or may not be sensitive features on the land (e.g., historic, archaeological, biological).** Existence of endangered species, known archaeological or historic sites, or areas with local covenants often preclude the conversion of agricultural land to certain selected land uses. In many cases, these restrictions apply not only to the sites themselves but also to adjacent parcels of land. The locations of such sensitive sites are exclusively spatial.

• **Any parcel of land has neighboring parcels, just as you have neighbors in your residential neighborhood.** In several cases, you have seen how adjacency to some types of land or to services (e.g., sensitive sites, roads, electricity) has an impact on decisions regarding agricultural land conversion. This is a larger issue, however, because many municipalities try to cluster certain land use types, including agriculture. It is generally considered more viable to have industrial parks rather than have industries spread throughout a community. Alternatively, agriculture adjacent to agriculture is thought to be more conducive for continued agricultural activity. As you might guess, the concept of adjacency is both explicitly spatial and is linked by measured distance to all other land uses and distance-sensitive features in the study area. In short, this will require you to derive the measure of adjacency through distance measurement, buffers, and so on.

• **Given that there are neighboring parcels, some land uses are mutually compatible and some are not** (Figures 6.7 and 6.8). Another factor that must be derived through measurement is that of compatibility versus incompatibility of adjacent and nearby uses. Doing this requires that you obtain a matrix of compatible versus incompatible uses. What seems on the surface to be a single factor is actually compounded by the number of possible uses of the land. In some cases, a generic incompatibility-of-uses map may suffice in the absence of a complete list of potential uses. In others, the model may have to be recalculated each time a new use is proposed.

Each of these basic factors relates to the goals of agricultural land preservation and planning either as potential thematic factors or as algorithmic operators for your model. As yet, however, they exist as a rather haphazard, possibly incomplete, and loosely connected set of concepts and factors. One way to organize them is to group them so that you can examine each group individually to evaluate them for completeness or redundancy. In some cases, it might prove useful to create the groups

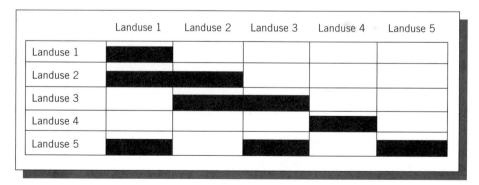

Figure 6.7 Land use compatibility matrix. One method of reviewing the compatibility of land uses, or any other factors you might be modeling, is to create such a simple matrix. In this case, the shaded compartments show land uses in columns that are compatible with others in the corresponding rows.

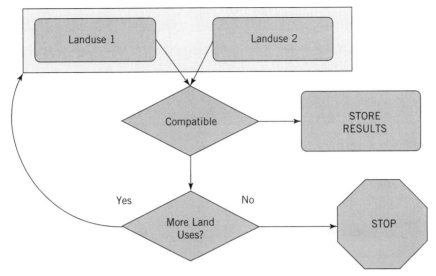

Figure 6.8   Flowchart for deciding on the compatibility of any two land uses. Each compatible pair is stored until all land uses have been examined.

before a list is compiled. This requires that you already know what the major planning compartments are. In some cases when the detailed factors of a model are not well known, especially when the geography of a model is not completely researched and assumptions are dangerous to model outcome, such a methodology is preferred. Models based on data mining are examples of this, particularly when you are using an explicitly inductive GIS modeling strategy.

GIS models of wildlife habitat use and prediction frequently fall into this category. Beyond being inductive, they are also descriptive, again because they allocate specific prescriptions not for where wildlife should be but rather where they exist. In this case, you are trying to predict where a selected species or group of species is likely to be found on the landscape. You will not always know all of the requirements of the species, so compiling a list of factors is more difficult than for our LESA model. However, you do know some basic things about wild animals. For this example, I will use the North American cougar or mountain lion. What general things do you know about cougars without searching through the detailed literature?

As with all mammals, cougars require food, water, a place to sleep, and a place to rear their young. The locations of some of these things may be identical, as with a den site. So you begin by creating the large compartments first. Below is your first cut at this compartmentalization:

- **Food:** Cougars are carnivores (meat eaters) who require a prey base. This prey base can be small animals like rabbits and squirrels or larger animals like deer and antelope. Assuming you do not know specifically what animals the cougar eats, you make an assumption that it fits somewhere in the rabbit to deer size range. Your initial reaction would be to identify all prey consumed by the cougar, and in what proportion. This may require years of research, examining remains of species for identification and cataloging, as well as extensive statistical analysis. Your purpose is to convert this knowledge into some map of where the cougar's preferred food lives. Now you have created a very difficult task for yourself because you must create an individual GIS model for potential locations of each of its prey species. In most scenarios, what is needed is a rapid-assessment methodology, deliverable in a realtively short time frame. Thus, you may have to wait to obtain such detailed information. Additionally, male cougars tend to travel greater distances for food than do females. It is obvious then, that your original compartment of food is really two compartments: *food for males* and *food for females.* Although it may be the same food, the distance from den sites is greater for the males, allowing you to look farther afield for potential prey sites. And, when females are nursing, they are much more likely to stay closer to their den when they search for food than at other times. You now see that your new second component can be further subdivided into *nursing female* versus *non-nursing female* models of prey availability.

For this grouping, you need to consider also that prey will try to avoid their predators as well, so it is not unexpected that you would find cougars in places where they can hide prior to attacking. Some cougars, such as the Florida panther, live in locales that lack terrain features and so will use vegetation with vertical structure in which to hide, whereas the mountain lions in mountainous parts of the southwestern United States will likely use such features as rock outcrops, ravines, and arroyos for hunting.

- **Water:** All mammals require water, including cougars. But whereas open water may be essential for nursing females, it is less important for males, who obtain much of their water from fluids contained in their prey. So you see that water may be included in the model, but it may require you to break your conceptual model into *water for males* and *water for females.* And, as before, your compartment for females may also have to be split into *nursing females* versus *nonnursing females.*

- **Dens:** Dens provide places for shelter from the elements and hiding places for avoiding detection. It is reasonable to assume that the den sites probably occur in selected portions of the terrain where either the rocks or associated vegetation provide hiding places. Additionally, some rock formations tend to encourage the formation of caves, but other do not.

In the case of the mountain lion model, you have already created the general compartments from which you can extract more detail as you learn more about the cougar. The GIS itself is likely to be of assistance here as you employ telemetry, collection of location evidence such as scats (excrement and claw marks, for example). By employing a site-specific analysis of each of these points, you can begin to understand what types of vegetation the cougar prefers to travel in, which in turn can be used to determine something about their prey base preferences for vegetation, as well as information about locations of water and den sites. In other words, you need to work backward to take coarse thematic data such as hydrology, surface geology, vegetation, telemetry, and scat data and derive more specific factors and themes as you go along. This type of modeling is difficult, and, as you have already seen, often requires iterative testing and evaluation to refine it.

HIERARCHICAL COMPARTMENTALIZATION

In the case of the cougar habitat model, we began by creating the coarse compartments relating to a fairly general, hierarchical knowledge of large feline predators. This was necessary because the model was primarily inductive and because our specific knowledge of cougar habitat requirements was both rudimentary and incomplete. We return now to the LESA planning model, where we began with a preliminary list of probable factors relating to our goal of agricultural land preservation. We need to find a way to compile a hierarchical grouping of these factors to assist us in identifying missing, redundant, and connected factors.

We have seen that there are some fairly obvious conceptual relationships that relate to the goals and objectives of our LESA model. These are derived from common sense, combined with a general working knowledge about how cities and farms work. For learning how the model conceptualization process works, we have chosen to work with a relatively easy to understand model. For those requiring more complex or more specific domain knowledge, it is best to have the domain experts working with the GIS modeler so they can share their ideas. But whether the model is a simple one requiring only basic concepts or a complex one requiring quite specific knowledge, a simple listing of potential geographic conditions, factors, and interactions likely to be involved in the model is not nearly enough. Instead, we need to systematize these original components into some ordered structure.

One simple way to do this is to create a number of categories that can be thought of as formal sets, each containing some collection of the items we have identified. As with any other set, these sets of concepts may share common elements. This approach allows us to group our common ideas without being forced to make formal linkages from one concept to another. It also allows us to have a first look at our categories to identify missing items in each set, to move them from one group to another, to eliminate unnecessary sets, or to add entirely new ones. This will become more important as we begin to perform model formulation, flowcharting, and final implementation. But it also can play a role in the early conceptualization of the model as well.

Let's take the list of ideas we generated for the site assessment portion of the LESA model. Do any of these concepts share common elements that will allow us to categorize them? There are many ways to group these factors, depending on what we are going to use the data for, so we need to keep our goal in mind. Our purpose is to evaluate the land as sites for either continued agriculture or nonagricultural replacements. Although we are not restricted to these categories, Williams (1985) used the formal groupings established by the Douglas County Planning Department in consultation with the USDA. We will begin our discussion there so you can see the mechanics of this grouping. This will also allow us to refer to the published work for guidance on future projects.

The established groupings for the LESA model in Douglas County, Kansas, include the following:

1. Land use/agricultural

2. Agricultural economic viability

3. Land use regulations

4. Alternative locations

5. Compatibility of proposed use

6. Compatibility with adopted master plans

7. Infrastructure

The first grouping deals exclusively with the agricultural land use near the proposed site of agricultural land conversion. As we saw earlier, the likelihood that agriculture will remain in an area is often enhanced if adjacent or nearby land is also in agriculture. Group 2 examines agricultural economic viability, a determining factor in land conversion. After all, if the land is viable for agriculture, we are not as likely to see its conversion to other uses. The third group is a legal set of factors, controlled in large part by government bodies that place restrictions on the land for its use, whether agricultural or nonagricultural.

Because a common factor in land use decisions is based not just on whether the land is capable of being used for either agriculture or nonagricultural endeavors but also on the availability of alternative sites for nonagricultural uses. This is the reasoning behind factor group 4, alternative sites. The availability of sites may be a function of zoning restrictions, or it may be that there are actually better places to put to nonagricultural uses. Urban expansion also places pressure on agricultural land resources, thus making the availability of land even more important.

Issues of compatibility are the foci for the fifth group of factors that attempt to link proposed nonagricultural uses with both aesthetic and physical relationships of the existing land with those uses that are proposed. These issues answer questions about whether a proposed use would interfere with the proper functioning of surrounding uses, disturb drainage patterns resulting in flooding, contribute to increased pollution, detract from the visual appeal of scenic areas, or disturb protected archaeological, historical, or biologically protected sites nearby.

Although there are legal restrictions on land, as well as demands on a limited resource base, many communities have controlled-growth policies in place that are reflective of a desire to manage land use change. These policies are most often a set of guidelines rather than restrictions and are frequently formalized as comprehensive plans. Many comprehensive plans exist as a set of statements describing the overall goals and objectives of the plans, but some are formalized as cartographic documents that suggest future growth zones. In the latter case, we can readily obtain spatial formalisms of otherwise less than spatial decisions.

The factor groupings just outlined are those originally designed by the Douglas County LESA working group and illustrate a fairly logical set of factors. As a modeler, you might ask whether these are the only groupings possible and whether a different grouping would create a fundamentally different set of outcomes. The answer to the first question is no, this is not the only way in which a GIS LESA model could be compartmentalized. And the answer to the second question is yes, sort of. There are many combinations and permutations available for even this simple descriptive model, and the way in which we choose to conceptualize our model will have an impact on the model outcome. One can assume, however, that, although there will be some differences from one model conceptualization to another, the results should not be wildly different if the model conceptualization was arrived at through logical approaches and with the same goals in mind. Luckey and DeMers (1986–1987), for example, compiled a grouping of factors for the same Douglas County mode that is somewhat different from the original, composed of only five groupings: land use issues, compatibility with adopted plans, other compatibilities, agricultural viability, and urban infrastructure. This was designed to reduce the redundancy of factors already identified by the working group. Other methods used the same groupings but derived slightly different sets of factors (DeMers, 1985).

To anticipate your next question, the conceptualization of a model is only the most general first step in its development, and model formulation and flowcharting, which we will cover in the next chapter, will help us to keep any number of model concepts reasonably similar to one another. Ultimately, each modeler will have different ideas, a different background, and different biases. Some approaches may achieve nearly identical results although they may be more or less elegant in their construct, whereas others may seem quite similar in concept and achieve different results. The goal is to produce a reasonable, verifiable, and acceptable model, especially a model that can be refined after examination of the SIPs. In short, the model should be both defensible and modifiable.

ADDING THE SPATIAL DIMENSION

In our discussion of the compartmentalization of a GIS model, we sometimes made reference to how we might move from factors to spatial data. Adding the spatial dimension to our conceptualized GIS model components is a matter of asking a typical geographer's question: Can it be mapped? This may seem simple to answer for those who are familiar with the vast array of cartographic representations and the dimensions and types of data that can be represented on maps, but many data types do not readily lend themselves to cartographic display, particularly because they do not have an explicit spatial dimension. Others have potentially spatial data that are not currently available in a cartographic format, such as tabulated statistics, point locations for cougars, and archaeological sites. Still others require the addition of some GIS analysis to form them into a useful set of thematic maps from which the model can be built. We will examine the LESA model to illustrate some of these problems and define some possible solutions.

Beginning with our first compartment or submodel as formalized by Williams (1985), we are interested in the existence of agricultural land in and around the proposed land conversion site. This rather vague statement actually has three basic components, all of which are fundamentally spatial: (1) land on site, (2) land adjacent to site, and (3) land within a specified distance of the site. Similar spatial components are readily available for at least some of our second group of factors—agricultural economic viability. As we saw earlier, the size of the farm—a fundamentally spatial factor—places restrictions on the ability of the farmer to do things, such as turn a tractor for tilling the soil and use large harvesting equipment. The size of the land parcels themselves, whether in agricultural production or not, also places restrictions on their potential agricultural viability because small parcels of land are less attractive as potential agricultural sites for the same reason that small farms composed of collections of parcels are. One factor suggested by the LESA working group is agrivestment in the area. This factor relates to monies spent on farm equipment, silos and farm buildings, roads, and a wide variety of nonspatial factors. In the research process, Williams (1985) was forced to abandon this factor, leaving a portion of the LESA model blank. This is certainly an option. We could either eliminate it from the model, thus suggesting that the LESA model itself be changed to reflect this omission, or keep the factor as a nonincluded component in the spatial version of the model. In the latter case, we might want to include a formal statement of noninclusion to document the relative incompleteness of the model.

Before we adopt this strategy, let us first examine more closely what possible spatial surrogates we might be able to employ to allow us to use agrivestment as a working component of our model. Let us assume, for example, that we have access to the financial records of the agricultural landowner. We could, for example, relate the total annual investment in farm machinery, new roads, new buildings, or improvements to any or all of these items. We could further divide these investments by the total farm size over which they are to be employed. This would give us at least one method of adding a spatial component to the model factor. Additionally, we could also incorporate a dollar-per-square-footage element of such factors as new buildings and roads that themselves take up space. This latter approach does not take into account the purchase or upkeep of farm machinery, but it gives us at least another way to include the factor in our spatial modeling. You might be able to think of some additional ways of converting this aspatial factor to some quantifiable spatial factor.

Another approach is to include the idea of agrivestment not as a spatial factor but rather as a nonspatial multiplier or operator. When, for example, we need to combine spatial factors concerning agricultural viability, we may be able to link them by an average yearly investment dollar amount for each farm. In this way, the agrivestment component is included in the modeling process, but without the need for explicit spatial data to support it.

Our land use regulations component is composed of three basic factors, two dealing with legal mandates related specifically to zoning and one that is not. In the first place, our spatial component comprises the availability of land zoned for specific nonagricultural purposes. To take a simple approach, we can assume that all land not zoned for agriculture would be eligible for conversion. This means that our spatial component is readily available from zoning maps. We could also modify our spatial component to include each of the possible nonagricultural uses separately. In this way, we could examine our zoning maps for light industrial uses, if that is the use to which our agricultural land is proposed to be converted. This might require us to run the model each time a particular land use conversion type is suggested. A second spatial component deals with alternative, off-site locations for the proposed use and would be obtained from the same data source—zoning.

The remaining land use regulation factor—the need for additional urban land—is somewhat more complex in that it requires us to create a spatial dimension to an initially aspatial factor. In this case, there are at least two fundamental parts to the spatial dimension that need to be examined. The first is the currently occupied urban limits, which would entail mapping or obtaining a map of the existing city limits. This would need to be modified, however, because the city limits represent not a homogeneous region but rather a fragmented region composed of a wide variety of existing land uses. As you can see, this requires us to create a subcomponent that combines these two concepts into a single group (Figure 6.1). The most difficult task in assigning spatial dimensions to alternative locations is in the wording, which states that we are looking for "availability" of alternative land and "need" for urban land. Quantifying these in the spatial dimension may take its simplest form in that we must total only the availability or need within our study area. Each of these then would be modified to represent a single-valued grid or thematic map based on the sum of land area. This works from a modeling perspective, but we might want to revisit this in more detail to determine if the approach is reasonable from a planning outlook. We would ask ourselves the following questions, for example: (1) Does a summation of existing urban land give us any indication of the past growth of the city? (2) Does a summation of existing urban land give us any indication of where urban growth pressure is currently greatest or where it is likely to be greatest in the future? (3) Does it suggest any knowledge about population growth or projections? For our current example, we will take the simplest approach, but it might be interesting to see if some urban growth models might not assist us in this endeavor or if the diffusion of urban or other land use types could be explicitly included in the model.

The next compartment requiring us to create a spatial dimension deals with compatibility of the proposed use. Compatibility with the adjacent or surrounding area is conceptually nebulous and requires us to evaluate adjacency and measure distances from the proposed site. In its simplest form, it suggests that the compatibility issue is one of comparing the proposed nonagricultural use of a particular parcel (spatially explicit) with other land uses (also spatially explicit). The nebulous part deals again with terminology. Before we can formulate or flowchart this part of our model, we must first define our terms explicitly. Which land uses are compatible with one another? Each pairing would necessarily have to be evaluated before this could be done. To simplify this problem, we could state some pretty obvious ones, such as these: (1) identical land uses are compatible with each other, (2) similar land uses are more compatible near each other than dissimilar ones, and (3) the most compatible rating would normally be given to agricultural land uses adjacent to other agricultural land uses because our primary goal, as defined earlier, is to preserve, whenever possible, the best agricultural lands. Compatibility with unique topographic, historic, or groundcover features or unique scenic qualities can be separated out as two distinct subcompartments because they deal with both on-site and adjacent lands. For on-site quality comparisons, we can make simple comparisons of features that are already mapped or can be mapped from locational data. Adjacency can be determined easily by comparisons of the target site with its near-neighbor grid cells. This latter process is, however, somewhat simplistic and could be expanded (either now or as the model is refined) to include such tasks as buffering and view-shed analysis, in which such eyesores as factory smokestacks and other aesthically displeasurable features could be sheltered from adjacent scenic land uses and features by planting vegetation or building topographic features that mask them. These would require a considerable amount of additional modeling and possibly be an added expense to the owner of the land who proposes the alternative land use. For our initial model, we can focus on the existing available spatial data, those revolving around existing and proposed land uses and unique features.

Another subcompartment will be necessary for evaluating compatibility with such activities that might impact flooding, drainage, or waste disposal and pollution problems. At a minimum, we will need soils maps and surface hydrology and drainage maps against which we can evaluate these problems. A simple approach to this problem is to compare the proposed land use with locations of soils prescribed for such use (soils maps). The assumption (not always a safe one) is that such maps have already taken such factors into account. The same can be said of maps of the 100-year flood zone. A more detailed analysis would require us to perform evaluations of surface and subsurface flow by using flow modeling capabilities of the GIS to ascertain potential water pollution and drainage problems off-site.

Our other major compartment dealing with compatibility is conceptually easier to establish as a spatial dimension: compatibility with adopted master plans and/or compatibility with areas assigned as designated growth areas. We could commonly assume that these are available in mapped form or can be transformed in a reasonable amount of time.

Our last compartment dealing with urban infrastructure requires us to obtain spatial data regarding each of the infrastructure components. Distances from city limits, transportation facilities, central water systems, and sewage, for example, would normally require us to measure distances (as buffers) from each of the known locations of these facilities and services. This is relatively straightforward as a coarse modeling task but can become more exact if we are examining the nodes along the linear features where hookups are most efficient and cost effective. Again, we can leave that for more advanced modelers.

IDENTIFYING POSSIBLE DATA ELEMENTS

Our last step on model conceptualization is to identify, with some basic level of specificity, the possible availability of spatial data. These data may not already be available in cartographic form, but they should be easily convertible to it. This is merely an explicit formalization of the last step whereby we added the spatial dimension to our initial conceptual model. Although identifying the actual cartographic elements is important, it is equally important that noncartographic factors and elements be accounted for here. As we have seen, this may involve the use of spatial surrogates to replace explicitly spatial data.

As an approach to identifying the spatial elements, we need to revisit our hierarchical compartments and treat each as a major branch in a tree structure (Figure 6.9). The final model can be thought of as the trunk of the tree, and each successive compartment, as smaller and smaller branches. As we get to the end of the branches, we have leaves, which are the actual map elements or source maps from which all other components are derived. This forces us to leave out the nonspatial components because they are neither branches nor leaves. We will, for the time being, ignore them and revisit them in the next chapter, which concerns model formulation and flowcharting. Here, we create our tree to identify if there are missing branches or leaves.

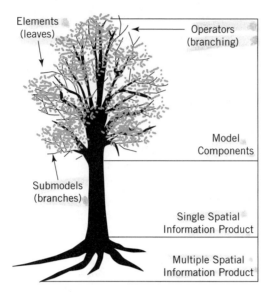

Figure 6.9   **C. Dana Tomlin's concept of a hierarchical model based on the analogy of a tree.**

When we review the conceptualization flowchart (Figure 6.1), we will note that it is not a purely linear one. Instead, we could view it as an iterative flowchart allowing us to answer yes and no questions at key points to help us along the way. By iteratively going through the conceptualization portion of the flowchart (the portion on the left), we continue to add or subtract factors on the basis of discussions with our client (or review of the modeling context if we are doing the model for ourselves). Theoretically, when we can answer yes every time to the spatial component question, we have completed the conceptualization. This does not, however, indicate that all of our factors are accounted for as explicitly spatial. It means only that we have either found spatial dimensions for those that we could, found surrogates for some that we could not, allocated others to operators rather than thematic cartographic data, or decided to eliminate them from the current model. These loose ends will be accounted for more explicitly in the next chapter.

Chapter Review

Conceptualizing a geographic information system (GIS) model is an extension of our spatial thinking applied to a specific set of GIS modeling goals and objectives. By first establishing our expected spatial information products, we work backward to identify the general spatial concepts and factors of which it is composed. For models for which we are relatively familiar with the general concepts or for which the concepts are relatively obvious, we can begin modeling by listing them and later compartmentalizing them to help us develop submodels. In other situations, particularly when we as modelers are relatively unfamiliar with the individual spatial factors involved, or when the purpose of modeling is to induce the spatial factor interactions, we create general compartments first, then fill them with more specific factors as we become aware of them. This awareness may be a result of the compartmentalization itself, or we may need to perform some form of iterative experimental or statistical procedures to assist us.

The Land Evaluation and Site Assessment (LESA) GIS model for Douglas County, Kansas, illustrates how identification of factors and related concepts can be an effective method of conceptualizing how the model will ultimately be put together. However, even relatively straightforward models like this require us to perform an iterative examination of the concepts to identify missing or redundant factors. By creating functional compartments, we can better view the overall model design and ascertain the completeness and integrity of our model. Alternatively, our mountain lion habitat model requires us to view broad-scale factors of habitat requirements, thus allowing us to refine the model through iterative application of experimental and statistical techniques applied to sample mountain lion data.

Once we are satisfied that we have a reasonable set of model factors, we must then determine which of these has readily available spatial components that we can map. Some factors are available as mapped data (e.g., zoning maps). Others will become available only as interim GIS modeling thematic maps resulting from GIS analysis (e.g., nearness to available sources of water will require us to perform some distance measurements). Some factors are inherently aspatial and will require us to develop spatial surrogates (e.g., need for additional land for nonagricultural uses). There will also be some factors that are either aspatial or for which there are no available data, which may require us to omit them from our initial model, pending such factors as future research, cartographic production, and legal or political decisions.

The conceptualization of a GIS model is meant only as a first step in the modeling process. When it is complete, we should have some general ideas of what thematic maps we might employ or create, generally how they might relate to one another, and their relative availability. What it does not do is to provide us with specific thematic map elements, operational linkages, functional sequencing, and specific GIS commands necessary to complete the model. That is the topic of the next chapter on GIS model formulation and flowcharting.

Discussion Topics

1. When we examined goals definition for descriptive geographic information system (GIS) models, we had two completely different different approaches: In the one example, we compiled a list of factors and concepts relating to our overall goals for the model, and in the other, we began by compartmentalizing first, then attempting to break these out into individual concepts and factors. Discuss the primary reasons for why we used these different approaches. What were the specific reasons? Could you argue that one method is superior to the other? Could either approach be applied to either model conceptualization? What do you see as limitations of one method over the other for selected model types?

2. What do you see as the fundamental differences between conceptualization of descriptive models and of prescriptive models?

3. A common task of GIS analysis is the description of suitable sites for such activities as solid waste disposal. As a group activity, select a portion of your immediate area surrounding the city. Divide the group into subgroups and begin the process of model conceptualization. First, decide if you are going to begin with a set of criteria or if you are going to begin by creating a hierarchical organization. Go through the remainder of the process described in Figure 6.1 until you define a stable set of spatial variables and criteria with which you can feel comfortable. Once finished, compare your results with those of the other subgroups. Do you see any differences? What accounts for them? Can you put together a larger, more comprehensive version of your model conceptualization after this discussion?

4. In our discussion of spatial components for the Douglas County, Kansas, LESA (Land Evaluation and Site Assessment) model, we discovered that the agrivestment category was essentially aspatial. We examined a couple of alternative

approaches to either eliminating this factor, creating an operator out of it, or creating surrogates. Discuss some alternative solutions and how you might be able to implement them.

5. Discuss the limitations of the LESA model when availability of alternative locations is the primary topic. How could population growth models, diffusion models, past and present urban expansion models, and urban growth directionality models be added to the existing LESA model to improve its performance?

6. What sophisticated modeling types might be needed to more accurately address the problems of drainage beyond the application of a simple flood zone map? Answer the same question for problems relating to on-site waste disposal beyond simply looking at the quality of the soils for such analysis.

7. Review the LESA site assessment factors that were examined in this chapter in light of the alternative grouping proposed by Luckey and DeMers (1986–1987) and with specific regard for factor interactions. Discuss the problem of duplicate factors, especially given that the land evaluation portion of the model is based on soils, as are some of the other site-assessment factors. Suggest ways that you might be able to remove this redundancy.

8. Discuss the advantages of a simple, linear GIS LESA model as opposed to a more complex one from the perspective of both model verification and model validation. Keep notes on your discussion for later discussions in Chapter 9.

9. Why is our conceptualization approach a linear one? Should we not have everything ready for our model once we have compartmentalized it and added the spatial dimension?

Learning Activities

1. Examine the list of factors that were developed in this chapter when we created a conceptual model for the LESA (Land Evaluation and Site Assessment) model. Now, compare these with those in T. H. L. Williams's (1985) article. Keeping in mind that Williams worked exclusively from the actual list provided by the Douglas County, Kansas, LESA working group, suggest what factors you would include in the LESA model for Douglas County. What does this tell you about how creating a list of factors prior to creating the database can enhance the model and its potential as a decision support tool?

2. Go to the following Web site (www.wiley.com/college/geog/demers314234/) and click on *Student,* then on *Online Resources for Students.* In the new window, go to the lefthand column under *Exercises in GIS.* A list of databases used for *Fundamentals of Geographic Information Systems* appears. At the bottom is the LESA database used by Williams and modified to run under ESRI's Spatial Analyst software. Download this database and examine the thematic maps he included. Look at the databases specific to site assessment factors and examine the maps and their contents. From your answer to question one, suggest if there is a need for more coverages to complete the model we conceptualized in this chapter.

3. Suggest how surrogates could be used to complete the Douglas County database where there are missing coverages, especially when the coverages are missing because of their inherent aspatial nature (e.g., aesthetics).

4. Go into your backyard or any back lot near your school or office and select a small rectangular area, say, 10 × 10 meters. Place stakes at each corner and connect

them with twine. Obtain a box of toothpicks and attach a small, rectangular piece of paper to one end of each to make small flags. Now walk back and forth in a systemmatic fashion, placing a single toothpick next to each anthill you observe. Although you may not know what the actual species are, identify each by some number, which you will write on the flag. You can also keep a record of what each species looks like (rather than trying to classify it). Now, go through the process of conceptualizing an inductive descriptive GIS model of ant locations. What groups of factors might you look at? What individual factors should you examine within each group?

5. In Discussion Topics question 5 in this chapter, you were asked to examine some possible ways of making the alternative locations factors more spatially sensitive by looking, for example, at how urban pressures might be greater in one direction than another. You were also asked to consider incorporating some alternative sophisticated modeling approaches that incorporated both space and time. Look at the existing spatial LESA database for Douglas County (http://www.wiley.com/college/geog/demers314234/) and determine what additional data, if any, might be required to perform such models within your software. This will require you to survey the literature to find examples of these models and how they have been applied to planning. Provide an annotated bibliography of the references you find.

## CHAPTER 7

Model Formulation, Flowcharting, and Implementation

On completing this chapter and combining its contents with outside readings, research, and hands-on experiences, the student should be able to do the following:

1. Create a flowchart for individual modules of a formalized GIS model using both a manual method or any of the available flowchart computer programs

2. Use the flowcharting capabilities of GIS software to adapt flowcharts to that software

3. Integrate the individual module flowcharts into a composite flowchart of the GIS model; in doing so, illustrate how the direction of flow proceeds for model formulation versus model implementation

4. Discuss why flowcharting is so important to formulating GIS models, particularly with regard to isolating essential thematic map elements and linkages among them

5. Illustrate some different methods of flowcharting a GIS model and discuss the advantages and disadvantages of each with particular reference to user needs and documenting the decision process for later model verification and validation.

6. Illustrate the role of model flowcharting in identifying missing and redundant thematic map elements

7. Demonstrate how aspatial data and/or operators can be incorporated into a model flowchart

8. Describe how the GIS model flowchart can be used to expand or modify a model as new data are discovered, new knowledge is found, or new approaches are developed

9. Describe and illustrate how the GIS model flowchart can be incorporated into larger, more complex model flowcharts, thus demonstrating the modular approach to GIS modeling

10. Demonstrate the use of the flowcharting capabilities of your GIS software to implement a model

11. Explain the role of metadata in GIS model implementation

12. Provide solutions for models whose environmental constraints are either too tight or too loose

13. Use at least one automated utility to create metadata

MAKING SENSE OF THE CONCEPTUAL MODEL

As we saw when we looked at model conceptualization in Chapter 7, the process of GIS modeling is often a complex, multicomponent, iterative process, especially when working with prescriptive GIS models. In fact, the prescriptive GIS model demonstrates the most interactive modeling type, often requiring multiple iterations of flowcharting, interim map creation, analysis, and output. This is the pinnacle of the cyclical process of GIS, where we move from one subsystem (i.e., input, storage and editing, analysis, and output) to another at will.

In the past, the process of model flowcharting was tedious, often requiring the use of plastic flowchart templates or unfamiliar software. As you will see later in this chapter, the flowchart is an elegant method of structuring the compartments we envisioned in the conceptualization process, for identifiying the necessary elements, and determining the functional relationships among the themes. Unfortunately, the functional separation of the flowcharting process from the actual modeling implementation discourages its use. Even the simplest prescriptive GIS model will frequently require multiple iterations, and some descriptive models as well require the development of interim thematic grids that must be evaluated before the next step can proceed. In such circumstances, most modelers will be more inclined to proceed with the session, examining interim grids, making decisions, and moving on to the next step, than to stop and produce a new flowchart or modify an existing flowchart by hand or by using separate software. What typically results from such an ad hoc and disconnected approach is that the modeling is done on the fly, with little or no formalization and with no documentation for later model verification.

Fortunately, some modern GIS software currently contains flowcharting capabilities that are explicitily integrated with the modeling process. This allows modelers to use the flowchart interactively, thus structuring their thought processess and creating interim thematic maps and interim flowcharts, each of which is invaluable both to the process of modeling and to the verification and analysis of model acceptability. In this text, I will use two such systems, the ERDAS Imagine Spatial Modeler and ESRI's Model Builder, to demonstrate how this works and how it can be used to both enhance and document the GIS modeling process.

Where the conceptualization of a GIS model is a fairly generalized task, and an important one, the interrelated tasks of flowcharting, formulating, and implementing a GIS model require a substantially more specific and exacting methodology. During model conceptualization, we created general compartments. We must now generate actual submodels for each of these, often breaking them into still smaller models. For each of these, we must now also move from general analog thematic maps to specific digital data whenever possible, and from general aspatial data to either digital surrogates or to mathematical and logical operators when explicit spatial data are lacking. The transition is perhaps the most important part of modeling because it requires the modeler to link all the components (cartographic data, operators, and weightings) of the model from the themes to the submodels and from the branches to the actual operational linkages. It also requires anticipating and employing the iteration procedures of the GIS to complete a functional model. We will begin by first looking at the compartments and breaking them down into their thematic grids.

EXAMINING THE FLOWCHARTING PROGRAMS

There are many different raster GIS software packages available, and each has its own method of allowing the user to create models, but many of them lack a GUI specifically designed to model directly from a flowchart. Many of them use an algorithmic approach that, although functional, sometimes keeps the new user from immediate access to the power of the GIS. Because flowcharting is integral to the systematic formulation, construction, implementation, and documentation of GIS models, I will focus on two of the more common professional GIS packages that include the explicit flowchart methodology—the ERDAS Imagine Spatial Modeler (ERDAS 1999) and the ESRI Model Builder (Environmental Systems Research Institute 2000). This is meant as a demonstration of the concept rather than as an explicit endorsement of either product.

Before we examine these products, let us look at a typical GIS model flowchart produced manually to get a feel for the components. A model of a wild-turkey habitat, created as a part of a master's thesis (Zeff 1991) (Figure 7.1) shows each map element as a rectangle, connected by lines that often have a process or procedure written as text above or below the line to illustrate how the elements are connected. This is probably the simplest form of model flowchart, although certainly not the only kind (cf. DeMers 2000a). The simplicity of such a flowchart is useful because it reduces the model to its most basic components and operations. Both of the professional packages we examine next are somewhat, but not substantially, more complex than this. Their complexity is more a function of how the software actually implements the components and what types of components are to be included in the model, rather than in how the wild-turkey habitat model itself might differ functionally from its original.

Let us view the set of icons used for each of these pieces of software to see both their relative similarities and differences compared with that produced by Zeff (1991). We begin by examining the icon bar from the Spatial Modeler software (Figure 7.2). Ignoring the standard selection tool icon at the top and the three operational icons at the bottom, we are left with the icons that will ultimately be used in the formulation of the model. From the second row, reading left to right, we have the

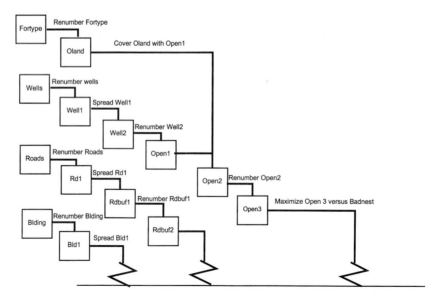

Figure 7.1 A portion of a simple wild-turkey habitat model. Although this is a simple approach to flowcharting, it is effective.

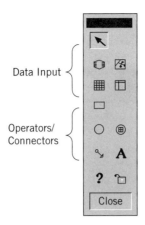

Figure 7.2 A portion of the ERDAS icon set for its modeling software. Notice how the data input icons are separated from the operators and connectors.

following: (1) raster input, (2) vector input, (3) matrix input, (4) table input, and (5) scalar input. Each of these represents a different form of element to be included in the model and each corresponds to the simpler rectangles from the Zeff (1991) flowchart. The remaining four icons represent the incorporation of (1) a function, (2) a criteria function, (3) a connector, and (4) an annotative text. Each of these, to one degree or another, is associated with the simpler lines and text used by Zeff in her manual flowcharts.

The flowcharting and formulation programs inside the ESRI Model Builder are simpler in appearance than those of Spatial Modeler but are no less representative of the general modeling framework designed by Tomlin. The opening icons for Model Builder are contained in a single horizontal strip that includes functionality for adding data, functions, text, and connectors and for adding and deleting bends (Figure 7.3). Bends allow for more elegant flowcharting. A simple implementation of the flowcharting program using the Douglas County, Kansas, LESA model as an example shows how we move from input data (watelines, in this case) through a process (buffering, in this case), resulting in a buffer map output (Figure 7.4). Notice how the input data are represented in the flowchart by a rectangle, the function is represented by an oval, and the output is represented as a rectangle with rounded corners. Although a bit different from the ERDAS product, the Model Builder GUI and its Spatial Modeler counterpart are both strikingly similar to the standard approaches to model formulation and flowcharting first envisioned by Tomlin. Each is designed to allow the user to both flowchart and document the formulation of the GIS model. This process is begun by isolating each of the submodels (compartments) and defining the specific data elements necessary to populate them and to solve their internal computations.

GIVING THE COMPARTMENTS GRIDS

We have used the tree analogy first suggested by Tomlin (1990) to describe the process of model conceptualization, and we will continue that analogy here. Look

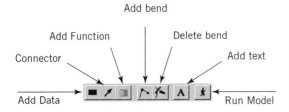

Figure 7.3 The ESRI raster modeling flowchart icons. These are relatively simple but provide all the necessary tools for creating a functioning model based on the flowchart itself.

Figure 7.4   A very simple flowchart using the ESRI Model Builder interface. This model shows how a buffer can be built around a gas line to create a buffer map. This would allow us to model the distance to other city services represented by linear data.

back at Figure 6.9 and its component parts. The trunk of the tree represents a single SIP—a rather rare occurance, but certainly possible. In the more common situation in which more than one output results from the model, the trunk can be ignored in favor of the roots. As we follow the trunk of the tree up to its major branches, each of these branches represents two things—submodels and the functional linkages among the submodels. Finally, the tree branches end in leaves, each of which represents the individual or elemental thematic grids representing the explicitly spatial data from our conceptual model.

Let us examine the Douglas County LESA model to get a feel for how we move from general compartments to the more specific model components and grid themes. To simplify our discussion, we begin by showing a list of components as they were originally defined by the Douglas County LESA working group (Table 7.1). We will further simplify things by continuing our model formulation, flowcharting, and implementation within some of the components of the same simplified model parameters defined in the Williams (1985) article.

You can see in Table 7.1 that we now have our seven compartments, each of which is meant to represent a portion of the overall site assessment system for LESA. Within these compartments, we previously defined some general possibilities for spatial data types and sources. The table lists, within the compartments or submodels, the source maps from which the final thematic map elements (the leaves of our modeling tree) used in the original GIS implementation were derived. Additionally, the table shows the relative weight of each of the additive site assessment factors. Although these seem straightforward, arriving at them requires that the modeler understands that functional linkages exist between the individual source maps and the site assessment factors they were meant to represent. In fact, one thing that has yet to be mentioned about the LESA site assessment factors is that each of the factors not only has its own weights but also has a degree of compliance or degree of conformity. This is the amount that each set of factors complies with the stated environmental conditions it is meant to represent. For example, for the Douglas County model, each of the following conditions or degrees of compliance is accompanied by an assigned score. Table 7.2 shows two of the site assessment factors, their degree of compliance with the environmental conditions, and some estimates of how these might be scored. These scores result in finalized grid cell values for each of the submodels (one for each factor). We can then combine these themes on the basis not only of their internal compliance scores but also of their weights, through a simple process of weighted overlay.

This is obviously a simple example. Other models will likely be more complex and involve a significant increase in computation. The concept remains unchainged, however. Each of the submodels has its own flowchart and its own model formulation. The formulation not only assists us in determining the necessary map elements to use but also helps us determine how its final outcome links to the other submodels or components. In our simple site assessment example, the linkages are simple weights or additions of one outcome thematic grid to another.

But before we leave this example, we need to look at it more closely because these seemingly simple weights are not directly available from their elemental grid themes. Taking the first factor of percentage area within 1.5 miles, for example, we see from Table 7.1 that we must break this down into three components: (1) areas with 95% of

TABLE 7.1 LESA Model Showing Relative Weights of Each Factor

	Source Map	Relative Weight
Site Assessment Factor		
Land use/agriculture		
1. Percent of area in agriculture within 1.5 miles	Land use	10
2. Land in agriculture adjacent to site	Land use	7
Agricultural economic viability		
3. Farm size	Parcel size	2
4. Average parcel size within 1 mile	Parcel size	4
5. Agrivestment in area	—	3
Land use regulations		
6. Percent of area zoned agriculture within 1.5 miles	Zoning	8
7. Zoning of site and adjacent to it	Zoning	6
Alternative locations		
8. Availability of land zoned for proposed use	Zoning	6
9. Availability of nonfarm and or less productive land	Zoning/soils	6
10. Need for additional compatible urban land	Land use/city limits	8
Compatibility of the proposed use		
11. Compatibility of proposed use with surroundings	Land use	7
12. Unique features or qualities	Unique areas	3
13. Adjacent to unique features or qualities	Unique areas	2
14. Site subject to flooding or in a drainage course	Surface hydrology	8
15. Suitability of soils for on-site waste disposal	Soils	5
Compatibility with adopted master plans		
16. Compatibility with an adopted comprehensive plan	Master plan	5
17. Within a designated growth area	Growth area	5
Urban infrastructure		
18. Distance from city limits	City limits	6
19. Distance from transportation	Transportation	5
20. Distance from central water	Water lines	4
21. Distance from sewage lines	Sewer lines	4
Total		114
Scale by 200/(114 × 10) = 0.175		0.175

the area within 1.5 miles in agriculture, (2) areas with 50% of the area within 1.5 miles in agriculture, and (3) areas with 10% of the area within 1.5 miles in agriculture. There are some definitions that must be expressed explicitly and some analysis that must take place to derive these values. First, we must define what the score will be for compliances of 50% agriculture within 1.5 miles. We must turn to the working group to assign a value here. A simple interpolation could easily be performed to assign a score of 5 or 6, depending on the desires of the working group. Next, we have to explain what

TABLE 7.2 Example Internal Compliance Values for Two of the LESA Factors

Factor	Compliance	Value
Percent agricultural land within 1.5 miles	63%–100%	10
	35%–62%	—
	0%–34%	0
Adjacent agricultural land	All sides of site in agriculture	10
	1 side adjacent to nonagricultural land	—
	2 sides adjacent to nonagricultural land	—
	3 sides adjacent to nonagricultural land	—
	4 site surrounded by nonagricultural land	0
...	...	...

Each compliance level has associated with it a value that can be assigned to each grid cell.

we do with the missing percentages (i.e., 51% to 94%, 96% and greater, and so on). This again requires us to interact with the working group to interpolate these values. We might easily expect that a reasonable solution would be as follows: (1) 95% to 100% area in agriculture within 1.5 miles would be weighted a 10, (2) 50% to 94% area in agriculture within 1.5 miles would be weighted a different value, (3) 11% to 49% would be assigned a lower value, and (4) 10% or less would be given a value of 0.

You should note from what we just did that we actually have four rather than five categories of compliance for this factor. This means we need to return to our original model and reconfigure it to conform to our new categorization. The outcome of this simple demonstration should begin to illustrate the complexity existing in seemingly simple procedures and models.

Assuming that we have settled on four compliance levels for our factor and that the working group is comfortable with the assigned values, we still have another question to answer: How do we derive these numbers? First of all, we don't have a site yet. This problem indicates that there is a potential for this model to move toward a more prescriptive type than anticipated. One way to avoid this is to examine the percentage of agriculture within 1.5 miles of every grid cell. This is a rather brute-force approach, but it works. For pedagogical reasons, we will assume we do have a proposed site and we are going to determine the percentage of area within 1.5 miles of that site. This requires us to perform two functions. First, we must create a buffer of a 1.5-mile radius around our proposed site. Next, within that buffer we must total the percentage of grid cells that are in agriculture. Finally, we must reclassify these grid cells according to our predetermined compliance values. Thus, we have a three-stage process for creating a simple submodel for this one factor. A flowchart of this submodel might look something like Figure 7.4. We continue this process for each of our factors, just as we would for any model's component parts. Each is a separate model with its own compartments. Our next step is to provide the linkages between compartments, just as we have provided linkages between the individual steps of the submodel.

LINKING THE COMPARTMENTS

Although we have seen how simple elements within our compartments (submodels) might be connected, this does not always explain how the submodels might be linked. This is partly because the submodels are themselves derived data and therefore are more complex than their elemental grid theme counterparts. In many GIS models, this is among the more difficult tasks because it requires a level of knowledge that does not always exist. The LESA GIS model we have examined at length several times demonstrates an explicit structure with equally explicit linkages among compartments. These linkages are simple additions or weighted overlay operations that are artifacts of a planning process that is itself very simple. But most models of real-world settings have neither explicit mathematical or logical linkages among the components, nor are they often understood well enough for such linkages to be easily defined. In most cases, we must make do with the best available knowledge and simple heuristics to decide how best to link one component to another.

Soil scientists have long used the Jenny (1941) general model of soil formation that suggests that soil is some function of five soil-forming processes. In this way, we can define soil as a mathematical equation, such as the following:

soil = f (climate, organic matter, relief, parent material, and time of development)

This could be thought of as a GIS model in which the output is a soil type, whereas the components are the five soil-forming processes. Unfortunately, these processes are themselves ill defined, their relative contributions to the overall soil condition unknown, and their interactions both complex and poorly understood. But most GIS models are not often either as poorly articulated as this example or as easily defined as our LESA example. Instead, most have a mixture of well-defined linkages and less well defined linkages. As is always the case with modeling, the idea is not to recreate reality but to use simplifying assumptions to generalize more complex real-world systems. As more is learned about how the systems, their linkages, and the individual components operate, the model can be revisited and refined to reflect a more realistic scenario.

There are many ways to connect compartments or submodels, just as there are many ways to connect individual components of a single submodel. However, because we are physically linking cartographic data from one grid to another—that is, because we are making direct comparisons with pairwise comparisons of each of the two sets of grids in each theme—this takes the form of one of the many processes of cartographic overlay. In some cases, the comparisons of themes may entail only neighborhoods of grid cells that are compared, as we saw in our examination of zonal functions. The zone that is applied to another grid will typically be a neighborhood of grid cells from another theme.

Among the best approaches to perform such overlays, no matter which of the overlay types is applied, is some form of local operator. As you remember, local operators work on a cell-by-cell comparison. This approach makes for models that are much easier to comprehend, understand, and verify later on. In particular, local operators make it easier to track any error levels as map data interact. They are also the easiest to explain to clients, often adding a higher degree of model acceptability, especially among clients who are not familiar with GIS modeling. Although many constraints in modeling approaches can be effectively handled by linking the submodels in this way, they will still often require more complex operators, such as focal, zonal, or block types of operators, within the submodels themselves. By performing these more complex operations within the submodels, then linking the submodels with local operators, we can embed the more complex aspects of the models in a general flowchart. This allows the modeler to explain the general model to a client in simple terms while de-emphasizing the details of the more complex internal operations of the submodels.

The most difficult to document and most computationally extensive operators are global operators. They are, however, some of the most useful for combining surface and flow data, functional distance operations in which one theme is used as a friction surface through which movement must take place. As before, however, it is better if these global operations are contained within the submodels to allow for easier documentation and acceptance by clients.

Although these simple guidelines give some general ideas about how to link your submodels, ultimately the selection of how they are linked is largely controlled by the functional relationships among the maps derived from the submodels themselves. As a rule of thumb, we have seen that simplifying the interactions among submodels is both easier to document and explain. Even if your client is GIS savvy or if you are the only one who will view and use the model, the submodels should be sufficiently well developed so that the output from each can be linked easily to others, especially if it can be linked through local operators. From a model-formulating perspective, this means that the submodels should be small enough so that they can be understood and so that they contain as few complex operations as possible. Then, if a model needs to be modified to satisfy changes in constraints, advances in knowledge, or iteration necessary for prescriptive modeling, it will not require reworking large portions of the formulation or the flowchart.

IDENTIFYING MISSING/REDUNDANT/CONFUSED THEMES IN FLOWCHARTS

As with linking compartments within the formulation and flowcharting of GIS models, creating compartments or submodels that have a limited number of computational operations will also simplify the identification of missing or redundant thematic data. There is no single rule of thumb for performing such identifications, because each model is unique. Most often, however, the flowchart itself provides substantial visualization, allowing the modeler to see where the leaves of our flowcharting tree are missing or where whole submodels might be lacking. This is especially useful when working groups are involved in the identification of relevant modeling parts. Even for individuals, however, it is often just a matter of examining our flowchart to see where we have branches but cannot identify precisely how the branch might be pieced together. In some cases, this indicates a lack of understanding of the way the system being modeled works, but in others, it simply shows that we know generally what must be modeled but do not have the necessary elemental thematic data to do it.

Identification of redundant themes can also use the flowchart as a tool for visualizing this problem. Redundancy will most often appear where elemental grid themes are identical from one model part to another. Although there are many situations in which elemental grids will be used multiple times, such as the use of a DEM to formulate slope ranges, azimuthal ranges, and visibility regions within a single model, redundancy will also appear here most often. In the LESA model we have examined throughout several chapters of this book, we have the opportunity for such redundancy because the land evaluation portion of the model is based on soil series as a basic element, as are some of the site assessment factors (DeMers 1985, Luckey and DeMers 1986–1987).

Although using the flowchart will, in many cases, be sufficent for identifying problems with factors, the modeler is not limited to this approach. Everything from mathematical and logical examinations to leveled data flow diagrams and from literature reviews to personal interviews have been applied. One technique that may also prove useful is to use a matrix of factors and its resulting consequences to the model (DeMers 1985). This technique is often applied in the development of environmental impact statements.

A modification of this last technique has also proved useful in identifying confusing environmental factors and themes. One example from the LESA model was the use of a factor interaction matrix (DeMers 1985) that illustrated how the seemingly straightforward site assessment factor "size of site or farm" resulted in confusion when the working group was asked what the impact of increasing the weight of this factor would have on the relative importance of the other factors to the model. As discussion arose from this confusing issue, the working group identified that the factor itself was fundamentally two separate although related concepts (DeMers 1985). The size of the farm and the size of the site to be developed drive the model in two different directions. Hence, they had to be separated as individual factors.

ADDING DATA SURROGATES AND ASPATIAL OPERATORS

As we have seen, there are occasions when explicit spatial data are not available for building GIS models. In some cases, this is because the factor is not really a factor at all but rather a linkage between other explicitly spatial factors. Put differently, some missing data are functions used to manipulate combinations of thematic data. As we have already seen, the four types include local, focal, zonal, block, and global operators. Some examples come from older versions of the original Map Analysis Package, which included the ability to "map" an entire thematic grid to a single value. Such themes do not really represent actual thematic data but rather provide multipliers, overlay weighting values, or are even portions of a regression equation as was applied by Tomlin (1981). These examples are aspatial operators because their locations within the grid are irrelevant because all the numbers are identical. Newer raster GIS software normally does not require that a grid be created. Rather, a single number can be added, subtracted, multiplied, and so on, to each of the grid cell values throughout the target grid. Similar approaches can be applied to such operators as neighborhoods, roving windows, and other nonthematic environments. Although these have explicit locations, they do not represent spatial thematic data.

Spatial surrogates are used in the absence of the specific thematic map data preferred for our GIS model. For some, the terminology may be new, but we use spatial surrogates in many typical GIS models. Remotely sensed data stand as among the most obvious and most common applications. Digital imagery often takes on the appearance of real maps and is often called an image map, but its fundamental basic data elements are essentially components of electromagnetic radiation. When they are not used as raw biophysical data (Jensen 2000), we use them as surrogates for other environmental parameters, such as aboveground biomass, vegetation classifications, land cover and land use, and many more. Other surrogates include a count of houses as a surrogate for population, home values as surrogates for income levels within neighborhoods, or soil class as a surrogate for agricultural viability. To use these variables, we need to recall the application of related variables as one form of dasymetric mapping (DeMers 2000a), in which we can predict one variable on the basis of the existence of another. Limiting variables and density of parts, as detailed in DeMers (2000a), are also common ways of establishing surrogates for missing variables. The numbers and kinds of techniques are limited only by the GIS modeler's ingenuity and experience.

Regardless of which surrogate type used, it is highly recommended that its legitimacy be tested prior to implementation. This is most often done by performing some form of regression analysis. Whether linear or nonlinear, parametric or nonparametric, bivariate or multivariate, or even logistic, some form of predictive statistical prediction should be employed prior to the use of surrogates. In fact, maps of residuals from regression of spatially correlated variables can often suggest new variable relationships that are not normally visible (Thomas 1964).

IMPLEMENTATION

Reversing a Flowchart (Running the Model)

The flowchart used to formulate a GIS model most often begins with the final output or derived thematic data. Traditionally, to run a model formulation we must reverse the direction, in that the smallest compartments are implemented first, proceeding one section at a time until the final output is created. Although the essential methodology has not changed significantly, the GUIs used in professional software generates a set of macro commands that systematically run the model this way. An essential premise is that you have organized the flowchart in a proper hierarchical format (see section on hierarchy later). The definition of a GIS model explicitly states that the order and sequencing of the model is essential. If properly formulated and flowcharted, the model will run properly. ESRI's Model Builder software provides a good example of how this works. The software creates a set of Avenue Script code that reverses the flowchart, creates all the intermediate themes (grids), and generates output. A prerequisite for this process is that the model must first be saved. This allows the model to be modified later on through changes in the GUI. This and similar software have made the entire process of running GIS models and testing their results a much more efficient process than in the past. In addition, they provide an effective means of applying iterative processes, especially those used for prescriptive models.

Iteration (Refining the Model/Adding Intermediate Themes)

Whether descriptive or prescriptive, GIS models will frequently require refinement, addition of more detail when it becomes available, and correction of mistakes in logic or mathematics when they are discovered. Prescriptive models will, on the basis of their situational sensitivity, also require the user to run them as each new situation is encountered. Most modern GIS software has the capability of storing command sequences or modeling steps (see section on recordkeeping later). Although not necessary, this does allow for easier modification of the model by simply changing the internal components of the model implementation sequencing rather than having to start from scratch each time.

For iteration, we have already seen that modern GIS software also includes iteration statements ("do until" statements or "if then else" statements) that allow you to do repetitive tasks. They may also allow for the input of the modeler rather than just running until some condition exists. In most cases, this runs as before, but the condition response is to ask for input from the modeler. As we will see in the next section, it is often much easier to perform these operations on each of the submodels first before proceeding to the overall model.

Even noncommercial software includes a method of recording operations as they take place. Therefore, if your software does not allow you to create a series of explicit commands prior to modeling, it can record the manual operations you perform and keeps track of the sequence of these operations. In most cases, this is written out as some form of text file that you can refer to or that you can rewrite as a set of command sequences later on. Whenever possible, you should use one or more of these techniques. This does not negate your ability to test individual sequences prior to running submodels, but the techniques will save many steps when you must perform multiple tasks or have complex models.

Model Hierarchy (Compartmental Implementation)

We have already seen that GIS model flowcharts allow us to break large, complex problems into component parts. The degree of compartmentalization used in model formulation will have a profound impact on our ability to perform iterative or repetative tasks, to isolate missing or incorrect grid themes, to correct mistakes when they are discovered, to verify the model results, to explain the results to non-GIS users, and to assess the acceptance of the model. For these reasons, I recommend that GIS models be developed as a set of leveled flow diagrams not unlike those used in data flow diagrams. The flowcharting software contained in the GIS packages does not explicitly perform this leveled approach in which each submodel appears alternatively as a single rectangle or as a complete model itself. Despite this minor drawback, the usefulness of the GUI for creating, implementing, and testing individual submodels has been greatly improved.

One technique for quickly evaluating whether your model structure makes sense is to create a sketch of your flowchart, as you were introduced to in the previous chapter. We begin by building the hierarchy in purely graphical form, from the trunk of the tree to the primary branches to the smaller branches. For example, if our model is one designed to identify likely locations for finding a species of endangered wildlife—let us say some carnivorous mammal—we begin by creating an output rectangle called *habitat*. As we saw with our cougar habitat example earlier, we recognize that food, water, and dens are the primary branches of the model. Each of these branches can then be broken down into more detail as we move from coarse hierarchical levels to more detailed levels.

This has been discussed in detail at the conceptualizing level, but we also need to look at it in context with the implementation level. The idea is to define the modeling components for each of the submodels (in our example, food, water, and dens). These might be called *foodsites, watersites,* and *densites,* respectively, to give the submodel output descriptive names. The idea then is to implement each of these branches of our conceptual tree—each submodel—separately. This allows us to perform any iterations, adding and subtracting grids, and tightening or loosening constraints without affecting the final model. Once each of these submodels is tested for functionality (if not validity and acceptability), the pieces can finally be merged into the general model.

A Binary Maps Approach (Dealing with Complexity)

As modeling takes place, it is difficult to keep track of the many categories possible for each of the grids. This is especially true when many of the grids interact through overlay operations and other comparative measures. Although the professional software will usually keep track of these, it may not always be possible to revisit all of these categorical interactions when modelers are asked to demonstrate how the model actually worked. Two simple techniques have traditionally been applied to assist in this. The first is to give each map (whether an input map or an output map) a name that is descriptive of what it represents. This is especially important for intermediate maps whose contribution to the model may be vital but whose existence is often ignored after the model is complete.

The second technique is related to the first in that it requires us, whenever possible, to limit our categories within each grid (again, particularly with regard to intermediate grids). A technique that many modelers use is to create, whenever possible, binary maps with category names such as *goodsoil/badsoil, goodzone/badzone,* and *goodslope/badslope*. In the early days of GIS, this was fairly commonplace, but the ability of the GIS software to deal with multiple categories has resulted in is relative disuse. It proved to be an effective method of recordkeeping and helped in model justification later on. There are many situations when such purely Boolean approaches are inappropriate—particularly when models require ranking. In such situations, ranked categories can also have descriptive terms that will assist modelers both in recordkeeping and in model iterations when they become necessary.

Recordkeeping (Maintaining Intermediate Grids)

In the past, the maintainence of intermediate or interim coverages was absolutely essential to allow modelers to explain how the model was actually constructed and implemented. This was especially true if no formal flowcharts (especially those showing the processes involved in the creation of intermediate thematic grids) were developed. But even in the absence of such intermediate grids, most raster GIS software, even the simplified, educational variety such as OSU-MAP-for-the-PC, contained procedures allowing for the storage of sets of macro commands that would allow for their recreation.

More advanced GIS software links the flowcharting and formulating software with the generation of such macro commands. Figure 7.5 illustrates the ERDAS Spatial Modeler flowchart for a focal analysis model. Using the software's model librarian and its Spatial Modeler Language (SML) (not to be confused with the PC version of ArcInfo's Small Macro Language (SML), we can generate the following command sequence:

```
# INPUT RASTER
# OUTPUT RASTER
# Focal Analysis
# FUNCTION DEFINITION
# Median Value
# Neighborhood Definition
#
# set cell size for the model
#
SET CELLSIZE MIN;
#
# set window for the model
#
SET WINDOW UNION;
#
# set area of interest for the model
#
SET AOI NONE;
#
# declarations
#
Integer RASTER n1_1ndem FILE OLD NEAREST NEIGHBOR AOI NONE
"$IMAGINE_HOME/examples/lndem.img";
Integer RASTER n3_MedianImage FILE DELETE_IF_EXISTING USEALL ATHEMATIC 16
BIT UNSIGNED INTEGER "$IMAGINE_HOME/examples/MedianImage.img";
INTEGER MATRIX n7_Low_Pass;
{
#
# load matrix n7_Low_Pass
#
n7_Low_Pass = MATRIX(3, 3:
        1, 1, 1,
        1, 1, 1,
        1, 1, 1);
```

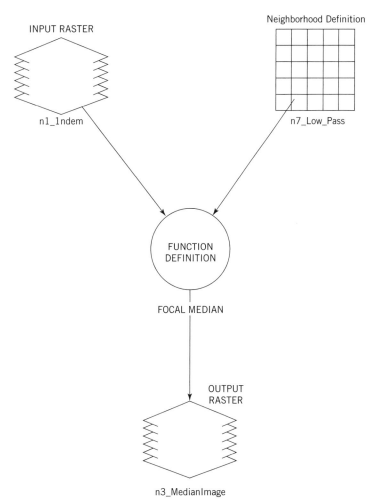

Figure 7.5 An example flowchart using the ERDAS Spatial Modeler software. In this case, the model is a focal mean operation.

```
#
# function definitions
#
n3_MedianImage = FOCAL MEDIAN ($n1_lndem, $n7_Low_Pass);
}
QUIT;
```

Essentially, this shows that the flowcharting program is acting as a GUI, whereas the actual code is being generated in the background. Although this does not explicitly store the actual intermediate themes, it does provide a way to recreate them if you desire. It is still a good idea to maintain the intermediate themes once the model is run, however, because this will allow you to run the model again and to compare the themes to assess any possible computational inaccuracies. We will examine this in more detail in Chapter 9.

Documenting Our Work beyond the Flowchart (Metadata)

Before GIS modeling became commonplace, it was not often necessary to document the sources, quality, categories, lineage, and other factors of the data that were used. This is primarily because there were so few models and the models were often academic exercises to demonstrate the ability of the early software to implement them. With the nearly logarithmic growth in the availability of professional GIS software, digital databases, and models using both, we need to be able to improve the interoperability of the models we create. This is particularly true where submodels created with standardized data sets for one government agency might be linked to a larger model in another such agency. For this reason, and many more, the U.S. government created the Federal Government Digital Data Committee (1992), which has adopted the SDTS for use by all organizations that provide data to the federal government. Among the most important tasks for such providers is to document as much of the important detail of the data as would be deemed necessary to permit other agencies to use the data for their own purposes.

Although many people view metadata as important in data set creation, they do not always connect the idea of metadata and data dictionaries with the process of modeling itself, except inasmuch as they will need to examine the accuracy of the model when it is complete. Although this is certainly true, many categorical descriptions that lack sufficient detail may result in a complete inability to run a model at all. For example, descriptions of such categories as *desert*, *wetlands*, *parkland*, and *early successional* are all frequently encountered in land cover and/or land use maps. Without substantial descriptions of what these mean, the ability to transfer the data sets from one model to another may be hampered or stopped completely.

Because of the potential impact on modeling, particularly with regard to model interoperability, it is necessary to include data dictionary and metadata data, as well as detailed flowcharts and/or macro command lists. Fortunately, some of these tasks are also becoming part of the model flowcharting and formulation process. Both of the professional raster-based GIS software packages mentioned here provide opportunities for documenting the conditions of use, important aspects of input in intermediate maps, and even details of functions used in the model (e.g., details about buffer sizes and numbers). Figure 7.6 illustrates the capabilities of the ERDAS Model Builder product for documenting both the grids and the modeling process as the formulation and flowcharting progresses.

These tools are becoming increasingly powerful, but you should also remember that the process of metadata production can become quite complex. Because of this, the annotation capabilities of raster GIS software are often insufficient to provide a complete, formalized, and FGDC-acceptable set of documentation. The following URL is one of the most comprehensive of an increasing number of Web sites that provide both automated and manual methods of creating metadata: http://www.fgdc.gov/metadata/toollist/metatool.html
Keep in mind that Web pages and Web sites change rapidly, so you may have to update your shortcut to the above URL from time to time.

Producing Effective Results (Tightening and Loosening Constraints)

Situations will sometimes occur in which the environmental conditions for any of the compartments, or for the model as a whole, either are met throughout the entire study area or cannot be met anywhere on the map. Both of these situations cause potential problems for the modeler. When the conditions are met for every cell in the

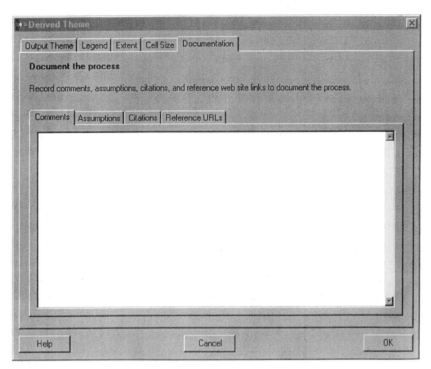

Figure 7.6   A portion of the user interface for ESRI's Model Builder software that shows how grids can be documented with data dictionary and other metadata information.

grid theme within a single compartment, the result is that you have gained no insights into your problem from that derived theme. In other words, the result does not contribute to the overall model. This suggests that the compartment might be eliminated from the model. Normally, this is not a good idea, because there was obviously some reason to include the compartment in the first case. This situation usually occurs when the numerical or logical constraints assigned to the compartment are too lax. The simplest solution to this, without eliminating the compartment entirely, is to ask the experts how best to tighten the constraints. Once the constraints are tightened, you must document this change in your data dictionary; then you will be free to implement the newly modified model.

There is the possibility, of course, that each compartment's constraints are tight enough so that at least some portion of the grid theme does not meet the criteria, but where, collectively, the final model grid fails to eliminate enough of the grid cells to provide a useful answer. Once again, this may require a tightening of constraints. In this case, however, it becomes necessary to evaluate whether the constraints on the individual thematic grids need to be tightened or whether the weighting of the themes themselves must be modified so that those grid themes that are more likely to constrain the model will become more important. Because each model is unique and the acceptable constraints are determined in response to the needs of the user, there is no simple solution to which of these approaches you should take. I might suggest that a sensitivity analysis be performed on the model grid themes to assist in determining a methodology. This sensitivity analysis need not be computationally extensive. In fact, it need not be mathematical at all. Often, the modeler has a feel for the importance of individual grids to the model and its effects on possible constraints.

But although circumstances arise in which the environmental constraints are so loose that the model fails to eliminate a substantial portion of the map because of its unacceptable conditions, the opposite is much more likely to occur. It is relatively easy to become too aggressive with the application of environmental constraints. In this circumstance, the outcome for either a compartment's derived themes or the general model's final themes is a set of maps in which none of the grid cells meet the environmental criteria. The idea is to loosen the constraints applied either to each grid theme or to some sensitive set of the thematic grid data used in the model. As before, constraint manipulation is best done in consultation with the experts and the users. Remember to include the new constraint levels and/or grid weights to your data dictionary and metadata.

Whenever the tightening or loosening of constraints takes place, there exists the possibility that the original intent of the models may become compromised. Care should be taken to ensure that this impact is minimized. Moreover, it is essential that any changes to constraints be well documented and that their potential impacts on the modeling results be explained, especially to non-GIS clients. Even if your model is meant for only you, it is important to keep both the final constraints values and the originals, perhaps as part of an extended data dictionary or as metadata so that you will remember how the model was formulated. This allows you to document not only what worked but also what was less than successful. The documentation of unsuccessful experiences is often at least as valuable as that of successful ones.

Chapter Review

Formulating and flowcharting geographic information system (GIS) models are critical factors in the modeling process. Today's professional GIS software frequently includes flowcharting programs that link these two processes with model implementation. This explicit linkage allows the modeler to stay focused on the process of modeling while making changes to the flowchart. Using the hierarchical or tree model, the processes of formulation and flowcharting move from the trunk (or roots, if there are multiple-output spatial information products).

GIS flowcharts allow modelers to visualize the overall processes that they attempt to mimic. By examining these documents, we can identify the kinds of elemental grids necessary to populate the submodels. Once established, these submodels must be linked through a variety of the local, focal, zonal, block, and global operators available to us in the raster GIS tool kit. This necessitates at least a basic understanding of the functional linkages among the submodels. In some cases, such as the additive LESA (Land Evaluation and Site Assessment) model, this may be preordained by the agency or organization responsible for its conception. For less formalized situations of natural systems and for socioeconomic systems, this is considerably more difficult. Prescriptive models introduce still more complexity. The flowcharts and alternative techniques of visualization and description can be applied to the selection of the necessary grids to model even the most complex models.

Moreover, these techniques can also assist in the linkage of submodels to each other; in the identification of missing, redundant, and confusing model components; and in the creation of spatial surrogates for those thematic grids for which they do not immediately exist. These same techniques can also support the selection of aspatial operators that link internal submodel components as well as the submodels themselves.

Because modern GIS software allows us to link the formulation, flowcharting, and implementation processes, we also have the capabilities of building basic models, one submodel at a time, and refining them as necessary. For more complex models, we can also perform iterative processing as new data are gained, as new concepts are discovered, or as scenarios change for prescriptive models. Finally, the flowchart, as a representation of the model formulation, will, because of the advances in GIS flow-charting capabilities, allow us to document the model as well as the descriptive components so necessary to development of metadata and for later verification, validation, and acceptance analysis.

Discussion Topics

1. Discuss why it is so important, particularly with complex geographic information system (GIS) models, to use flowcharting and formulation prior to implementing the model.

2. Why are integrated flowcharting programs more useful than either manual or separate flowcharting programs for formulating your GIS models? Include in your discussion the role of these integrated flowcharting programs in presenting a final model to a client and in verification and analysis of model acceptability.

3. From the literature, select an article that describes the flowcharting and formulation of a GIS model. If the article contains a flowchart, try to define any missing pieces not described in the text supporting the flowchart. If it does not have a flowchart, try to create one from the description of the model. Discuss where there are pieces missing and explain why you think they are missing and how you might fill in the blanks.

4. Jenny's (1941) five soil-forming factors can be thought of as five submodels for describing how soils are created. They are described in your text as an example of a coarse GIS model. As a mental exercise, devise a plan for creating a GIS model that results in a prediction of soil type based on the five soil-forming factors of climate, organic material, relief, parent material, and time. Formulate and flowchart this hypothetical model. Suggest other such loosely defined ideas as potential GIS models.

5. Just for the sake of argument, let us assume that you are attempting to implement the GIS version of the LESA (Land Evaluation and Site Assessment) model and that while doing so, you discover that when you examine the sensitive areas of Douglas County, they occur throughout the county. In fact, they occur to such a degree that this particular submodel suggests that no land can be converted. Discuss how you might loosen the constraints on this factor. As you do so, also discuss how this impacts the model results and the overall purpose for which the LESA model was developed.

6. The GIS LESA site assessment discussed here and in previous chapters suggests that the submodules or compartments are relatively simple. Using the discussion from this chapter and the following list of site assessment factors and compliance levels, describe in detail the unanswered questions and missing pieces necessary to flowchart and formulate complete submodels.

Site Assessment Factors, Their Weights, and Their Compliance Values

Percent area in agriculture within 1.5 miles (weight 8)

Value	Condition
10	95% of area in agriculture
—	50% of area in agriculture
1	10% of area in agriculture

Land in agriculture adjacent to site (weight 10)

Value	Condition
10	All sides of site in agriculture
—	One side of site adjacent to nonagricultural land
—	Two sides of site adjacent to nonagricultural land
—	Three sides of site adjacent to nonagricultural land
1	Site surrounded by nonagricultural land

Size of farm (weight 7)

Value	Condition
10	≥ 120 acres
—	80–120 acres
—	40–80 acres
—	20–40 acres
—	10–20 acres
0	< 10 acres

Average parcel size within 1 mile of site (weight 9)

Value	Condition
10	≥ 120 acres
—	80–120 acres
—	40–80 acres
—	20–40 acres
—	10–20 acres
0	< 10 acres

Agrivestment in real property improvements within 2 miles (weight 10)

Value	Condition
10	High level of investment in farm facilities (long term)
—	Moderate level of investment
0	Diminishing level of investment

Percent of land zoned agriculture within 1.5 miles (weight 8)

Value	Condition
10	≥ 90%
—	75%–89%
—	50%–74%
—	25%–49%
1	< 25%

Zoning of the site and adjacent to it (weight 6)

Value	Condition
10	Site and all surrounding sides zoned for agricultural uses
—	Site zoned agricultural and one side zoned low-density residential
—	Site zoned agricultural and two sides zoned for residential, commercial, or industrial
1	Site surrounded by residential, commercial, or industrial zoning

Availability of land zoned for proposed use (weight 6)

Value	Condition
10	Undeveloped land zoned for proposed use is beyond the primary and suburban growth areas of the incorporated cities
0	No zoned land available for proposed use (this value can be assigned only when a parcel is within the primary or suburban growth areas)

Availability of nonfarm land or less productive land as alternative site within area (weight 6)

Value	Condition
10	Large amount available
—	Moderate amount available
0	None available

Need for additional urban land (weight 8)

Value	Condition
10	Vacant, buildable land within city limits, capable of accommodating proposed use
1	Little or no vacant land remaining within the city limits to accommodate the proposed use

Compatibility of proposed use with the surrounding area (weight 7)

Value	Condition
10	Not compatible—high-intensity uses
—	Somewhat compatible but not totally
0	Compatible

Unique topographic, historical, or groundcover features or unique scenic qualities (weight 3)

Value	Condition
10	All of the site
—	Part of the site
0	None of the site

Adjacent to land with unique topographic, historical, or groundcover features or unique scenic qualities (weight 2)

Value	Condition
10	On all sides of the site
—	On three sides of the site
—	On two sides of the site
—	On one side of the site
0	None of the site is adjacent to these unique features

Site subject to flooding or in a drainage course (weight 8)

Value	Condition
10	All of the site
—	50% of the site
0	None of the site

Suitability of soils for on-site waste disposal (weight 5)

Value	Condition
10	All of the site
—	50% of the site
0	None of the site

Compatibility with an adopted comprehensive plan (weight 5)

Value	Condition
10	Soil limitation that restricts the use of septic system
—	Limitations to the soil can be overcome by special management
0	Little or no limitation

Within a designated growth area (weight 5)

Value	Condition
10	Rural area
—	Clinton reservoir sanitation zone
—	Suburban growth area
0	Primary growth area

Distance from city limits (weight 6)

Value	Condition
10	> 2 miles
—	≤ 2 miles
—	≤ 1.5 miles
—	≤ 1 mile or less
—	≤ 0.5 miles or less
0	Adjacent

Distance from transportation (weight 5)

Value	Condition
10	Limited transportation access dominated by rural township roads
—	Access to improved county roads or highway within suburban growth areas
—	Access to improved county roads or highway within primary growth areas
0	Access to full range of transportation services

Distance from central water system (weight 4)

Value	Condition
10	No water within 1 mile
—	Water within 2,000 feet
0	Water at the site

Distance from sewage lines (weight 4)

Value	Condition
10	No sewer line within 1.5 miles
—	Sewer line within 1 mile
—	Sewer line within 0.5 miles
0	Sewer line adjacent to site

Learning Activities ··

1. Select three of the seven groups of site assessment factors for the Douglas County, Kansas, geographic information (GIS) LESA (Land Evaluation and Site Assessment) model. Further divide them into their individual grid cell map elements. Now create a simple flowchart (using just boxes for data and arrows and text for linkages and functions) of each of these submodels.

2. Now, link the three submodels from Learning Activity 1 together to show how they are linked. Keep in mind that although this is a simple additive linear model, it also has an explicit weight assigned to each thematic grid. This weighting must also be included.

3. Now use the automated flowcharting capabilities of your raster GIS software (if available) and adapt each individual flowchart from Learning Activity 2 to that software.

4. Download the Douglas County, Kansas, LESA database and obtain a copy of the Williams (1985) article from which the data set was created. Use the flowcharting software of your GIS program to implement as much of the site assessment portion of the LESA model as you possibly can. You may have to make unilateral decisions about scores and specific details.

5. Using the model you created in Learning Activity 4, make some changes in your scores or in the weights that you assigned to submodels and run the model again to get a feel for the capabilities of the GIS to store models and allow for these adjustments.

6. Using the Douglas County LESA database, imagine that you are a federal agency and are planning on using the data set for purposes other than LESA. Now, using one of the available metadata generators or a printed version of a metadata questionnaire, list and define the aspects of the thematic grids that are poorly defined or completely missing. Describe how not having these details might present problems for using the database for alternative applications beyond LESA.

## CHAPTER 8

Conflict Resolution and Prescriptive Modeling

On completing this chapter and combining its contents with outside readings, research, and hands-on experiences, the student should be able to do the following:

1. Define some specific examples and types of spatial conflicts for which GIS could be used to help resolve them

2. Define the difference between site factors and situation factors with regard to spatial conflict resolution and land allocation

3. List some basic examples of using GIS to resolve spatial conflicts by displaying siting alternatives

4. Using available raster GIS software and existing databases, implement some basic form of spatial conflict resolution using the Orpheus Land Use Allocation Model discussed in this chapter

5. Define the strengths and limitations of the Orpheus Land Use Allocation Model for spatial conflict resolution

6. Explain how consensus building and hierarchical approaches are used in spatial conflict resolution

7. Describe how sensitivity analysis and principal component analysis can be used in spatial conflict resolution

8. Briefly describe how the displaced fuzzy ideal provides for spatial conflict resolution

9. Suggest additional research needed in the application of a GIS in resolving spatial conflict

10. Discuss the roles of model conceptualization, formulation, and flowcharting in spatial conflict resolution, especially where model validation is concerned

11. Explain why raster GIS has advantages over vector GIS for spatial conflict resolution

12. Discuss the role of the error component as it applies to spatial conflict resolution

162

INTRODUCTION

As we have seen, among the most important characteristics of GIS modeling is its ability to assist in making decisions concerning space (Cromley and Hanink 1999). Some of the most difficult decisions are those that concern conflicting demands on a limited land resource base. These conflicts go beyond border disputes, where ownership and precise locations are the primary issues, to prescriptive models of land allocation within borders, where stakeholders have opposing views about the use of the land resource base (Lesser et al. 1991). These latter problems are often exemplified by the LULU problem (locally unacceptable land use) or the NIMBY problem, when, for example, communities are opposed to the placement of waste treatment plants, solid waste facilities, high-tension wires, cellular phone towers, roads, and other public structures because of aesthetics, noise, pollution, odor, health concerns, economics, or other potentially negative impacts either perceived or real. Both the analytical and the output subsystems of the GIS can be of assistance in making decisions that either eliminate or lessen these impacts. In other cases, the software can provide verifiable means of ascertaining the legal rights and responsibilities of both parties or means for suggesting alternative sites that satisfy, more or less, both of their conflicting requirements.

This chapter will illustrate some straightforward methods of using the GIS to assist in the process of spatial conflict resolution and will briefly suggest some as yet unimplemented methodologies based on alternative logics or statistical techniques. In some cases, the GIS is used primarily as a visualization tool to display the sources and/or degree of conflict as a basis for discussion and consensus building, whereas in others the analytical tools are used to build prescriptive solutions based on suggested solutions. These latter two approaches are most often performed in an iterative methodology, where the stakeholders respond to scenarios, thus interacting directly with the GIS modeler. Finally, in the most complex case, the GIS can be used to derive alternative solutions, again through prescriptive modeling, that allow decision makers to finalize plans whose legal mandates (e.g., eminent domain) leave little opportunity for outside participation but where minimization of impacts is the final goal.

Although the GIS may seem an appropriate technology for multiobjective prescriptive modeling, primarily aimed at resolving conflicts, there has been surprisingly little published literature on the subject (Tomlin and Johnson 1991), although practical real-world examples are probably available but not published in the scientific literature. It seems that the majority of GIS modeling for decision support seems to be relegated to descriptive modeling, particularly to generate output from which public officials can make decisions. The output is most often in the form of cartographically presented rankings of possible sites for land uses. These, of course, are useful, but they illustrate a somewhat limited application of the technology. The relative lack of strongly prescriptive models is possibly due to their rather messy, difficult, and often iterative nature. Additionally, the prescriptive model most often requires frequent, intense, direct interaction with the decision makers during the process. Only recently has this aspect of GIS been addressed by the GIS research community. Perhaps this chapter will inspire some collaborative efforts between the research community and the decision makers using GIS in their work.

SPATIAL CONFLICTS

Spatial conflicts arise in many scenarios, particularly when human demands on the land interfere with its natural functioning, but also when one human use decreases value or suitability of another. Siting power line corridors involves many of these human–natural and human–human use interactions. When one is siting such corridors, there is nearly always a substantial traverse of land that must be crossed from one location to another. The strip of land allocated for power line corridors has the potential to interfere with existing drainage, creates an environmental edge that encourages the invasion of edge species or the exterpation of interior species, sometimes requires the allocation of lands already owned by others through the process of eminent domain, preempts other uses by subdividing the land, and involves many other conundrums.

In an urban/suburban setting, we have already seen the possibilities of conflict between agricultural and nonagricultural uses. Our discussion of the LESA GIS model showed a wide array of zoning, aesthetic, compatibility, and infrastructure factors that must be addressed to adjust for the expansion of urban areas and the growth of business and industry within agricultural regions. Necessary but often unseemly activities such as surface mining, placement of cellular towers, and building dams and water diversion structures illustrate the ongoing potential for displacement of existing land uses, degradation of property values, and reduction of the aesthetic quality of other land uses.

These examples all refer to two interacting types of siting criteria, each of which must be addressed. The first of these, referred to as *site criteria*, deals primarily with the direct impact on the actual site being considered for land use change. In our LESA GIS model, we examined some of these site factors when we attempted to rate the viability of agriculture versus nonagricultural uses on the basis of, for example, the size of the parcel, its current zoning regulations, and whether it was within a flood zone.

The second set of siting criteria is called *situation criteria* and deals with the impact on the surrounding area. In other words, these criteria are more appropriately called *neighborhood criteria*. As you will see later in this chapter, the situation criteria are not as easily defined because they most often do not just deal with off-site impacts but also require us to know beforehand the particular land use for the site and its potential impact on the surrounding area (situation).

Moreover, it is not enough to examine situation factors for a particular land use, but we have two other important factors that need to be addressed. The first is just how far from the site the impact zone might be. Arbitrary or average distances have been applied to such impacts, as we saw in the LESA GIS model, where 1-mile and $1/_2$-mile distances were applied. These are often designated very conservatively but are also chosen on the basis of rather arbitrary decisions largely controlled by our own measurement system. In Europe, for example, a 1-kilometer or $1/_2$-kilometer distance might be just as commonly designated as is a 1-mile or $1/_2$-mile distance in the United States. The second important consideration in examining situation criteria is an explicit definition of what is being impacted. We need to define, for each potential land use conversion, our concerns, whether it be with the off-site impacts on hydrology, with the off-site impacts on wildlife, with the off-site impacts on land values, or with something else. Although these same considerations might just as easily be applied to site criteria, their situation impacts are often more difficult to define, more spatially extensive in their outcome, and more common in occurrence.

It is clear that given a nearly endless set of different land use types and an even larger set of possible site and situational impacts, performing an allocation of all of these becomes a rather daunting task. It is often necessary to define limits on both before doing such an allocation, particularly if all are to be accommodated at once. A first step is to compile a list of only those potential land uses that are to be considered. After all, not all land uses are common or even likely for any given site or situation. We find, for example, that considering rice patty farming in a desert environment is not a useful endeavor. Each region normally has a limited number of land uses and a small set that are potentially viable in a given region of the world.

Once the potential land uses are defined, a second list of environmental criteria must also be considered, each related to some general categories such as economic viability, aesthetics, and wildlife habitat. These can then be decomposed into smaller, quantifiable, and—whenever possible—mappable characters. When both of these levels are available, we can begin the descriptive modeling task of defining possible alternative locations for each land use under each general category of environmental criteria.

GENERATING ALTERNATIVES

As we saw earlier with our LESA model, the GIS is capable of generating alternative uses of the land resource base by rating each parcel of land on the basis of its capability (the ability of the land resource to support the use) and or even of its suitability (how well the use fits with a wide range of planning factors). As with the mandated LESA model, the National Forest Management Act (NFMA) of 1976 requires the use of planning models to integrate planning into management of land resources (Iverson 1986). More importantly, the act combines criteria with the National Environmental Policy Act of 1969, requiring that the planning accommodate multiple use and environmental analysis in its framework. The result was the National Forest Planning Model (FORPLAN), a linear programming model designed to address these needs. Although it is not, strictly speaking, a GIS modeling approach, its elements are similar to those one might employ for multicriteria decision making, especially where land uses conflict. More importantly, it stands as a model that could be made explicitly spatial through the use of GIS (Carver 1991).

The idea is one of generating constraints on the use of the land for particular land uses, much as we saw in our descriptive model of the site assessment factors of the Douglas County GIS-based LESA model. Unlike the LESA model example, however, where we were primarily concerned with agricultural versus nonagricultural uses and where we often treated all nonagricultural uses alike, the idea of generating alternatives for each possible land use type is more explicit. Ideally, each possible land use type is analyzed on the basis of its ability to satisfy a set of criteria. These criteria can be Boolean or they can be weighted, as we saw in the LESA model example.

The results of constraints modeling are most often either a set of maps showing areas that cannot support particular land uses (especially in the Boolean approach) or rankings of possible areas for each land use type. As a decision tool, such maps are invaluable. They provide a set of possible solutions to land use allocation and a set of alternatives. As a basis for discussion, they illustrate clearly where limitations on the land exist, thus preventing particular uses. However, if one were to overlay these maps, it would quickly become obvious that many land uses are viable in certain areas of any region. Level, fertile soils may very well support nearly any type of land use. Although we can, as we saw earlier, tighten the constraints, thus placing certain limits on some uses, we still are left with competition for the use of some land parcels. In short, we have not provided the decision makers with optimal land allocations.

THE ORPHEUS LAND USE ALLOCATION MODEL

In earlier chapters, we defined the differences between descriptive and prescriptive GIS modeling, whereby the prescriptive model goes beyond a description of what is or even what could be and moves into more decision-based questions of what should be located where. Essentially, the prescriptive model allocates particular uses to individual parcels to achieve the best possible solution. Most often, this situation occurs when more than one possible land use can appropriately reside in a particular land parcel. In situations like this, the GIS becomes more than a device for showing alternatives. Among the best approaches to conflict resolution is an iterative one developed by Tomlin and Johnston (1991).

The Orpheus project was specifically designed to facilitate the land use allocation process beyond the display of constraints and land use alternatives. In our discussion of the LESA model, one of our criteria was the compatibility of the proposed use with a comprehensive master plan. It is precisely this type of master plan that the Orpheus project was designed for. Using a 35-square-mile study area some 50 miles west of Chicago, the authors performed a series of cartographic transformations and combinations of existing base data. The land use allocation model moved from siting criteria and constraints mapping, as we saw earlier, to a complete general master plan.

Descriptive Component

As with virtually all prescriptive models, the Orpheus approach begins with a descriptive component designed to illustrate the potential for each of 16 individual land uses within the site: conservation, agriculture, forestry, mining, urban and recreation, water supply, solid waste disposal, professional offices, manufacturing, retail shopping, religious institutions, housing, roads, cemeteries, and a research and development complex (Tomlin and Johnson 1991). Each descriptive submodel represented for each land parcel the constraints imposed on each of these 16 land use types, describing how particular on-site (site) and near-site (situation) environmental qualities such as production costs or historic preservation reflect either the existing conditions, such as soil properties or demographic composition, or could be made to exist, given the proposed use. The idea is to include multiple accounts where each might have an impact on site or situation (Brown et al. 1994).

As an example, the authors suggest a submodel for siting a housing development where the primary environmental concern is its impact on wildlife habitat quality. The descriptive submodel defines house siting criteria by illustrating the cause-and-effect relationships between housing development and wildlife habitat. As with all of the other descriptive GIS submodels, this one examines the site criteria relating the proposed land use and the currently existing site conditions. Additionally, each describes the situation criteria relating the potential interaction of the proposed land use with situations that will likely exist near site as a result of the designated land use.

For their housing development site criteria, for example, they suggest that one primary cause-and-effect relationship might be that a new house within a forest degrades the wildlife habitat more than one built on open land away from the forest. This type of criteria, again the site criteria, results in a ranking of suitability scores for each proposed land use, such as the housing development. The site criteria were, in this case, derived by employing a set of rules obtained through questionnaires eliciting expert opinion on what the important factors were, how they were important, and how important each is in terms of costs, environmental impacts, and effective land use (see Chapter 6).

The situation criteria for the proposed housing development are somewhat different in that they might indicate that an isolated single new house degrades habitat quality more than does a new house in the immediate vicinity of other houses. In other words, single houses have less impact on the surrounding habitat than does an entire housing development. Because this and other situation criteria often relate to impacts, they cannot be expressed initially as suitability maps. Instead, they will, in the first case, be expressed as a set of criteria or rules for the later development of such suitability maps once the proposed land uses do exist. These rules, applied to situational factors, also required that distance rules be developed, in consultation with experts, that could result in land use specific proximity maps outlining suitability on the basis of some maximum and minimum distances. If you review the factors of the LESA model, you will see the same types of distance metrics already in place, but not specific to each potential land use type.

The descriptive modeling would have resulted in a total of 16 suitability maps, one for each proposed land use. Each of these suitability maps is a descriptive model, and each is composed of at least two parts: a site suitability map and a situation suitability map. Each of these is based on a wide array of possible criteria and decision variables. These were necessary preparation for the prescriptive portion of the modeling by defining the nature and structure of the allocation problem.

Prescriptive Component

The prescriptive portion of the allocation model is designed to iteratively generate part of the solution to the allocation problem. Somewhat akin to mathematical optimization and some of the methods available in operations research, the purpose is to achieve, as closely as possible, individual allocation objectives. In the raster domain, one might think of each iteration as individual moves in chess or the Game of Eights or the Tower of Hanoi problem. Another analogy is the moves in solving the Rubik's Cube puzzle. In each of these situations, one attempts to solve a part of the puzzle. Once each step is performed, the puzzle situation is changed and new scenarios need to be produced.

As one might imagine, the combinations and permutations are nearly endless; however, each time a particular scenario is examined, we can suspend our disbelief long enough to assume that certain land use types actually do exist and are or will be put into place. Thus, our next iteration takes place with this assumption (Figure 8.1). Even under such circumstances, however, a perfect match satisfying all possible land uses and all possible land use conditions is not often possible. It is likely, however, that a degree of model stability will be achieved that satisfies most land uses and most environmental constraints. At this point, the modeling process can stop.

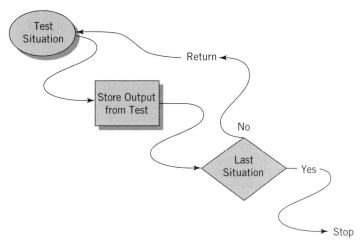

Figure 8.1   Flowchart illustrating how an iterative approach can be applied to situational settings in complex models. Each new situation is tested until you run out of situations.

The exact definition of an appropriate stopping point is not a simple one to derive. It will, of necessity require substantial and ongoing input by the stakeholders of each land use type, including nonanthropogenic types such as wildlife habitat. For both the descriptive and prescriptive modeling portions of the land allocation problem, methods of arriving at mutually acceptable solutions among the participants in the modeling process will have to be created. The literature on spatial conflict resolution is not robust, but there are a few relatively common techniques that have been applied, and we will discuss these now.

CONSENSUS BUILDING

Among the most easily implemented techniques for conflict resolution is the idea of consensus building. This is a method of arriving at generally agreed on terms, conditions, and limits. Consensus building is best begun during the descriptive modeling effort and continued throughout the modeling process, particularly during the iterative phase of the allocations themselves. The process of consensus building can be relatively unstructured but ideally should include output maps of each scenario for each allocation criteria whenever possible. Although this may require substantial inputs of time, it is more expedient to deal with land allocation problems and conflicts one at a time, as they arise, to avoid having to dissect the model components after substantial portions have been completed.

A more structured approach, also likely to include output maps at each critical phase in the allocation, includes the now common and time-tested Delphi technique developed by Rand Corporation. Although not commonly applied to the land allocation, especially where multiple objectives are involved, it has been used successfully under such circumstances (DeMers 1985).

HIERARCHICAL TECHNIQUES

Weighting and reweighting of factor variables is a consensus-building technique that has shown promise in the land allocation process (Davis 1981, Ive and Cocks 1983, 1989). The weights assigned allow conflicting demands to be placed in a form of hierarchy within which decisions can be made. This is a technique not unlike the original idea of assigning weights to the LESA model that we have already encountered. But unlike the LESA model, the weighting and reweighting are performed as an iterative process whereby the decision makers define the importance of individual factors and potential impacts as they are encountered. As with the other techniques already discussed, applying these approaches while the descriptive portion of the model is being developed is highly desirable.

Rather than making Boolean decisions regarding the acceptability or unacceptability of particular environmental factors or land uses, hierarchical techniques allow for compromise. Although the approach is not necessarily structured, the result is that each of the stakeholders is allowed the opportunity to give and take where, for example, one set of environmental constraints for a particular land use scenario can be assigned lower weights by a stakeholder given that other environmental constraints or conditions will be assigned higher weights under different circumstances and competing stakeholders will be asked to lower their priorities under the latter scenario.

A pairwise comparison of environmental conditions under different scenarios is one way of handling the complexity and interactions of factors often resulting in spatial conflict. As early as 1977, Hopkins recommended a hybrid technique of linear or nonlinear combinations of factors followed by rules of combination. Lyle and Stutz (1983) recommended a stepped matrix approach—also a hierarchical approach—to disaggregate each factor for each land use into a variety of causative factors. Its major drawbacks involved its oversimplification and the lack of understanding of feedback and factors interrelationships. One promising approach from the technology forecasting literature incorporated into land use planning within a GIS was the use of simulation modeling, independent of the GIS itself but prior to development of the descriptive model (DeMers 1985). The technique employed the Kane Simulation Model (Kane 1972) to isolate the potential impacts of one land use factor (whether site or situation) on other land use factors. It combined a modified Delphi technique to elicit expert opinion on pairwise combinations of factors so they could be incorporated into the GIS LESA model. The results, although promising, were aimed explicitly at the descriptive modeling portion of land allocation, and there is a need to test whether these or similar techniques could prove useful for the prescriptive portion of the model, especially where iterative examination of environmental factors and land use scenarios are applied.

STATISTICAL TECHNIQUES: CONTENT ANALYSIS

Although the most common approach to isolating important factors and deciding on how multiobjectives are to be addressed within a prescriptive GIS model is manual and interactive, we have seen that there are techniques such as the Delphi method that can assist us. By manipulating the numbers achieved through discussion, we can begin to automate some portions of the process, generally with the intent of making the resulting model more objective. There are a number of statistical techniques that can be applied to ascertaining related variables and identifying the most important concepts within a set of data. Some of these are already applied on a regular basis in remote sensing. Although it is not standard practice to apply statistical techniques for factor and factor weight determination, as well as conflict resolution, it is worth looking at them briefly for those who have a background in statistics and who may find it useful. I will suggest two basic methods that might readily be applied. You may have others you might want to consider once you see the potential for applying these.

One method that I might suggest involves the application of both qualitative and quantitative content analysis. In its simplest form, content analysis examines textual documents in digital form (word-processing documents). By breaking down the fundamental language construct, the software analyzes it for important descriptive words—a technique usually referred to as parsing. Next, it performs one or more methods of statistical cluster analysis or principle components analysis on these descriptors to determine which words used by the people who wrote the text tend to cluster into similar functional groups and/or to determine the importance of certain concepts.

Let us take an example using using our LESA model as a starting point. Let us say that we have the developers and the Douglas County, Kansas, planners write out a justification for why they need a particular parcel of land for either agricultural pursuits or nonagricultural activity. The text is read into the software for each participant. The software then uses a number of techniques, depending on the software chosen, to analyze common words and to compile common words into clusters or to decide which factors seem most important to each participant. We might find that land developers find that *distance to city limits* is a very important concern of theirs because of their ability to obtain ready access to *city services.* These may prove to be the very same factors that are important to the planners because they wish to allow *nonagricultural uses near the city*, thus allowing agriculture to take place farther away. This illustrates just one potential outcome that might provide important insights for further discussion and should allow a more rational method of selecting factors that are not readily available and for weighting which are more important to each participant.

There are many forms of content analysis and many software packages available for its application. Rather than suggest any here, I recommend you examine the Web page at the following URL for a robust list of available packages, sources for purchase, and their prices: www.gsu.edu/~wwwcom/content. Both qualitative and quantitative methods have the ability to provide useful information. Additionally, the application of the software requires the participants to examine their needs independent of one another. Furthermore, it forces each participant to prioritize those needs without undue pressure. Finally, it has the added function of removing the emotion from at least one phase of the negotiation process, thus allowing conflict resolution to take place. Although such techniques are currently experimental and there is little available literature suggesting its application, those modelers who are building real-world prototype systems, especially within an academic environment, might find the approach worth testing.

DISPLACED FUZZY IDEAL

A final note is warranted here about a potentially powerful although admittedly academic method for dealing with conflict resolution; it involves the application of fuzzy logic. Academic circles vary widely on their interest in, knowledge of, and acceptance of fuzzy sets and fuzzy logic. Let me first dispel the common myth that fuzzy sets and fuzzy logic are based on probabilities. They are not. Instead, they are based on an extension of traditional crisp or Boolean logic and set theory, where fuzzy logic allows a gradation between yes (1) and no (2). It is in the gradation that most confusion occurs. The distance between 1 and 0 is not a probability function that assumes a normal distribution and is based on the central limit theorem. Instead, it is a grade of set membership, typically assigned by estimates of where, within the yes and no space, it seems the best answer applies. In the context of our spatial modeling with GIS, one might use the degree of importance a LESA factor might have to an individual group participant in the development of prescriptive modeling. A developer might say, when asked, "Is distance to central sewage system important to you?" that it is very important. Alternatively, we could ask the participant to respond in the following manner: "Yes (0.9)." In this way, the participant is saying that it is a very important consideration in his or her business by assigning it the set membership value of 0.9, which is as close to an absolute 1.0 as possible without saying that it is an absolute necessity.

If you can imagine participants in a multiobjective GIS modeling working group all providing such responses, and if you assume that there is a robust mathematical environment in which to analyze such responses, then you have taken the necessary steps in identifying how they could be used for conflict resolution. Unfortunately, as with most conflicts, the reasons for conflict and the levels of conflict are not static and are subject to changes in tolerance levels, targets, and interactions with the other participants, which led Yee Leung (1988) to suggest that a method of identifying a nonstatic target through the application of fuzzy logic could provide a basis for conflict resolution. This dynamic target, which he calls the Displaced Fuzzy Ideal, has been attempted in the development of an expert systems shell designed for use in GIS (Leung and Leung 1993). Although it is not commercially available, to my knowledge, such a technique may prove to be a useful tool in the future and might suggest further experiments by advanced GIS modelers, particularly in the academic environment.

Chapter Review

Some of the most difficult conflicts to resolve are those involving multiobjective designs on a limited resource base. A geographic information system (GIS) is a useful tool for examining these potential conflicts by allowing the modeler to generate alternatives and to examine the potential impact of scenarios on other portions of the study area. This approach is predominantly employed in the development of prescriptive models in a multiobjective framework. A readily applicable general framework for examining conflicting demands and adjusting to them is called the Orpheus Land Use Allocation Model. Within this method, one first creates a descriptive model on which scenarios can be built. The descriptive model must be capable of describing the degree of compliance of two fundamentally different sets of environmental factors: site and situation. Site factors are those that directly impact the parcel of land being evaluated. Situation factors are those that are most often off-site factors, including adjacent parcels of land, and often require scenario building.

The scenario-building portion is the prescriptive portion of the model and may require that each potential situation be examined or modeled on an individual basis. When all scenarios have been examined, their impacts have been weighed, and spatial conflicts have been resolved through discussion with the participants in the model-building process, the final model can be completed. The Orpheus method suggests that through the display of the individual results of situations, discussions among the participants will readily result in resolution of most issues.

The process of resolving the spatial conflicts is not a simple matter of display of scenarios, however, and several techniques are suggested as methods of accomplishing this. Consensus building is among the most commonly applied and time-tested methods and includes the use of the Delphi technique for obtaining participant views. Alternatively, hierarchical techniques, such as weighting and reweighting, have also been applied successfully. Statistical techniques, particularly those employing qualitative and quantitative content analysis of documents provided by the participants, may also provide useful insights. Finally, the application of the Displaced Fuzzy Ideal may show promise in the future as knowledge-based GISs become more commonplace.

Discussion Topics

1. List and describe several types of land use conflicts that commonly require some form of spatial conflict resolution. Include in your list whether the conflicts involve displacement of existing land use, devaluation of land values, reduction of aesthetic value, or detrimental environmental consequences.

2. Describe the difference between site criteria and situational criteria and provide some concrete examples of each for some of the conflicts you listed in topic 1 above.

3. Discuss the role of descriptive modeling in preparation for prescriptive allocation modeling with a geographic information system (GIS).

4. Explain the role of the output subsystem of the GIS in providing decision makers with the necessary tools to perform their job; also, illustrate the limitations of this descriptive approach to land use allocation.

5. The Orpheus Land Use Allocation Model provides a structured approach to dealing with conflicting demands by addressing each environmental factor (whether site or situation) for each proposed land use type. Its limitation is that it still requires the input of humans to provide the actual resolutions of spatial conflict. Describe some of the techniques discussed in this chapter that might be applied to this portion of Orpheus. Discuss the pluses and minuses of each as you see them.

6. No matter which conflict resolution technique is employed, the solutions to land allocation problems still rely on maps. Discuss the potential problems of land allocation when maps of poor quality are part of the decision-making process.

7. Discuss the role of hierarchical methods such as factor weighting in reducing or resolving spatial conflict within the land allocation problem.

8. What potential role can fuzzy logic play in the resolution of spatial conflicts within knowledge-based GISs?

Learning Activities

1. As you have seen in this chapter, there are frequently situations in which conflict, especially spatial conflict, is virtually unavoidable. The types of situations in which this occurs are almost limitless, so we will restrict ourselves to a seemingly very simple case in which we have only to fit a number of land uses within a specified amount of available soils, each type of which has its own specific characteristics. This particular exercise is somewhat of a capstone activity in that it incorporates many of the topics covered not just in this chapter but in much of the GIS modeling process itself. As such, you should take some time on this. Focus not on the actual solution, but rather on the techniques that are applied to it.

Activity Learning Objectives

• To develop a plan for a selected nondeveloped region where the soils limit the uses of the land and where there are also space allocation requirements

• Because of the space and soils limitations and the multiple demands for the use of the land, there will be conflicting demands for the use of the land; determine a methodology for resolving the conflicts that arise as in planning for the land

Background

You are to assume the role of a junior partner in the firm of DeMers GIS Consultants, Ltd. (DGC). A housing development company has requested that GDC perform an analysis of the developmental potential for a site in Douglas County, Kansas. The parcel measures 50×60 grid cells, each of 2.5 acres, for a total of 7,500 acres of land. The development company has secured the purchase option on this property. They require the following distribution of land uses on this tract:

• 1,875 acres—single-family slab houses

• 938 acres—single-family houses with basements

- 937 acres—two-story multiple-family dwellings on slabs

- 469 acres—three-story multiple-family dwellings with basements

- 478 acres—sewage lagoon

- 938 acres—small shopping center (one-story slab buildings with parking lot)

- 937 acres—school, community center, city hall (all two-story buildings)

- 938 acres—parkland and recreation

The area is not served by a municipal sewage system. The sewage lagoon will have sufficient capacity to serve the shopping center and all other multiple-story buildings. The sewage lagoon should be built within one-quarter mile of all buildings that it will serve, to save money on sewer lines. You must consider the noxious odors generated by the lagoon and take into account the prevailing winds in this part of the country. Other constraints that you will have to consider are

- All single-family houses will use septic tanks.

- The sewage lagoon and shopping center must be on continuous tracts of land.

- The shopping center should be located at the edge of the parcel for highway access.

- All the land uses will require soils suitable for road construction, except the parkland.

- The parkland and all housing areas can be divided into any number of smaller parcels.

- The schools, community center, and city hall can be located on different parcels, but each requires a minimum of 20 acres.

Problem

As a DGC bright young consultant, you immediately secure a copy of the Douglas County soil survey from your Soil Conservation Service district office. In the back of the survey, you locate the area that the development company is interested in purchasing. (It just so happens that it is the Douglas County LESA study area we have discussed before.) The site has been reproduced as an ArcView Spatial Analyst grid theme in your LESA database. *You are strictly limited to the use of this one theme to solve the problem in this exercise.*

After having looked over the site, you consult the survey to determine what types of soil are in the area and how they will affect Stick House's development plans. You will, of course, read all about the soil type and become an instant expert on the soils in Douglas County. After having become said expert, you now consult the tables that have been developed by U.S. Department of Agriculture (USDA) soil scientists that rate each soil with respect to selected land use activities. The appropriate information from these tables has been reproduced for you below. You find that each soil has been given either a "slight," "moderate," or "severe" rating. The "slight" rating means that the desired land use activity may be pursued with little or no modification of the land. If the soil is given a "severe" rating, it means simply that you have to remake the world. However, do not be overly concerned with finding areas with a "slight" rating; with modern technology, all things are possible.

As you attack this problem, you will need to identify the soil limitations for each of the proposed land uses and renumber them as appropriate. You may have more than one land use identified as being best suited for a particular soil type. Some soils will be suited only for one type of land use; for example, the Kennebec soils are suited for use only as parkland and recreation. In this case, your job is simple and you can allocate that area to the single desired land use. When you have more than one land use suited for a particular soil, you will have to devise a strategy to decide which goes where. One approach is to look at the problem associated with the soil and determine if one type of land use is better suited than another owing to the type of problem associated with the soil. For example, if you have land that is equally suited to slab housing and playgrounds (let us say that the land is rated as having moderate problems due to flooding), then you might select the playground site over the housing because flooding would not be as critical a threat to the playground.

You will find that there are no ideal solutions to this problem. One area is better suited for some developmental activities than another, but many areas will have problems that will have to be circumvented using innovative building techniques or extensive modification of the existing landscape if Stick House is really interested in developing the area. That is not your concern. You are merely required to make your recommendations to Stick House, and you can let your conscience bother you all the way to the bank as you deposit your consulting fee.

Materials Needed

- 1 Douglas County soils theme in the LESA ArcView database (available on the Wiley Web site).

- 1 soil capability table

Information Products

As an output from your research, you are to provide a group report including

1. Eight implementation flowcharts, one for each of the individual land uses

2. A detailed description of the process or processes you employed to deal with the conflicts that arose among the land uses (e.g., did you have to relax or tighten constraints? require the developers to spend money to overcome existing constraints? do consensus building?). You may want to refer to your notes to discuss item 2.

3. One map showing your final results (hard copy or disk)

4. For a more difficult approach to conflict resolution, rebuild the LESA model using the same database but work from a prescriptive model, based on the Orpheus Land Use Allocation Model, that allocates each of the following nonagricultural land uses while still attempting to maintain the best agricultural lands for farming:

   a. Housing development

   b. Office buildings

   c. Green space (parks)

   d. Commercial development (a mall)

## CHAPTER 9

Model Verification, Validation, and Acceptability

On completing this chapter and combining its contents with outside readings, research, and hands-on experiences, the student should be able to do the following:

1. Define the differences among GIS model verification, validation, and acceptability analysis

2. Explain how the procedures of a GIS can be reversed to check whether the GIS software created the answers it is expected to produce

3. Describe the use of manually prototyping small areas to develop a set of correct GIS solutions to the larger GIS model

4. Describe some effective methods of determining the relative elegance and operational efficiency of multiple GIS models for the same thing (computational time versus flowchart elegance)

5. Discuss how to deal with missing data when surrogates and other solutions cannot be applied to the final outcome

6. Define methods of double-checking the logical or mathematical methods employed in a model to determine its validity

7. Create appropriate methodologies to examine specific algorithms employed in GIS models to determine how well the models represent reality

8. Create appropriate methodologies to examine the degree to which clients understand the output from analysis

9. Create and implement methodologies to define the degree to which clients accept and/or are able to implement the model for decision making

10. Describe the applications of statistical and spatial techniques for determining the validity of GIS model output

INTRODUCTION

Since the 1980s, there has been a great deal of research on GIS accuracy assessment and measurement. This continues a long-standing tradition of such research in remote sensing. In most GIS situations, this focus has concentrated primarily on the accuracy of individual coverages or themes, on the processes of input, and even on the interactions of the maps during processing. Although this is extremely important and very worthwhile, it often ignores the most imporant aspect of GIS modeling—that of the overall correctness and acceptability of the final model output. Remote sensing professionals like Jensen (2000), Lillesand and Kiefer (2000), and many others have made it quite clear that the final step in the remote sensing process is not the production of the output but rather the explanation to and acceptance of the results by the client.

As GIS modelers, it is essential that we take this same view of GIS output. The output is nothing more than a pretty picture if it does not exhibit correct implementations of the algorithms, if it does not substantially render the environmental processes being modeled, and if it is not an acceptable and useful product for the client. The output from GIS is potentially far more varied than that of remote sensing and can include noncartographic forms. For this chapter, I will limit the discussion to decision aids in traditional cartographic form. Although there is probably a need for more focused research on the assessment of alternative output types, most GIS applications and models still rely heavily on cartographic output. Perhaps these pages will encourage further research of this nature for both cartographic and noncartographic output. In the meanwhile, the following pages are meant to provide not detailed, time-consuming, research-oriented investigations but rather some more pragmatic general rules of thumb that can be applied to real-world settings.

DEFINITION OF TERMS

The process of GIS modeling has as its final goals both the effective modeling of the spatial phenomena under consideration and the utility of the results to the user. These are fundamentally different aspects. The terms most often employed in the analysis of GIS output are *verification*, *validation*, and *acceptance*. Depending on the context within which these terms are applied and the particular reference you use to define them (e.g., dictionary or thesaurus), the terms can often be used almost interchangeably. Their definitions are relevant here only because we need to know the topics we are examining. For our purposes, then, I provide some definitions with which we can proceed in our discussion. You are free to call them what you wish later on.

For this discourse, we will separate each of these terms as fundamentally different. At the same time, we will assume that both verification and validation are two aspects of the overall correctness of a model.

To verify a GIS model is to perform some method of determining if the model's computational code correctly represents how the algorithms were designed to operate and that they operate consistently on the same data set. The two fundamental questions are whether the software outputs correct numerical values when the coded algorithms are applied and whether such values are consistent from application to application when applied to the same data set. Although we generally assume that the computer software we use provides correct answers to questions posed to it, this is not always as true as one might think. Because the GIS algorithms programmed into the software rely on analog logics, mathematics, and calculus, we must make the further assumptions that these are correctly employed by the programmer and that they will provide the same answer in their digital equivalent that they would if they were performed manually. Such processes as converting differential equations to their algebraic difference equation equivalents, rounding error resulting from single versus double precision, and the ordering of mathematical and logical procedures all have the potential and unfortunate impact of making the results somewhat in variance under certain circumstances. Within this context, verification can also include an analysis of not just the correctness of the mathematical computation but also its level of precision (the consistency of results from multiple applications to a single data set). If an overlay is performed between two grid themes several times, the result should be identical. Or, if a buffer is created at 1 mile from a road network multiple times, the final buffer for each application to the same database should be consistently 1 mile, not 1.1 miles one time, 1.0 a second time, and 0.9 miles a third time.

Validation is often applied to acceptance and utility of the model as a decision tool. However, many dictionaries use the term *validation* as a definition of *verification.* To avoid confusion, we will use the term *validation* as a separate aspect of the same question. That question is simply: Is the model correct? In this case, the correctness is not so much a matter of its mathematical correctness but rather whether or not the algorithms represent what we intended to model in the real world. In other words, does the model provide a reasonable representation of the processes and spatial interactions of the real-world phenomena being examined?

Finally, acceptability, although operationally linked to both verification and validation, is used here to represent measures of a model's acceptability and utility as a decision-making tool. A model might be correctly identifying appropriate use of the land, but it may not be accepted by the planners who are responsible for using it in the performance of their job. Although an accurate model is more likely to be accepted, it is not a guarantee. Conversely, an incorrect model may be considered acceptable because it fundamentally represents the bias of the user.

As a modeler, it is generally more important that we concern ourselves with the correctness (verification and validation) of the model than with its acceptability. However, acceptability is still important if we are to fulfill the requirements of our users, especially if those users are ourselves. In this regard, model acceptability is not so much a matter of its correctness, or repeatability, but more often includes such aspects as ease of use, extensibility to other locations or environments, utility as a decision tool, understandability, flexibility for use with different scenarios, and even an ability to incorporate new knowledge or additional modules when available.

MODEL CORRECTNESS

Verification

Tomlin (1991) suggested that a fundamental process of model verification is a determination of the computational accuracy of the algorithms employed. Nearly analogous to performing multiplication to check your long division, the process requires that you know two things. First, you must be completely knowledgeable about how the algorithm or algorithms are actually meant to work. This is not always readily available when you are using professional GIS software, because many of the algorithms are proprietary. Second, you must also know the expected results beforehand. For very large areas, this can also be problematical. Both of these problems have solutions, however.

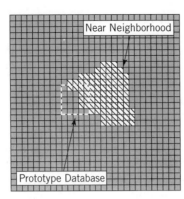

Figure 9.1 **Selecting a subset for evaluating correctness of focal function algorithms.**

For our first problem, that of knowing whether the GIS code is accurately implementing the prescribed algorithms correctly, it is not necessary that you examine this during the modeling process. It might be helpful to test this out well in advance of doing your modeling tasks. This generally amounts to a process of benchmarking your selected GIS prior to using it. But even though a complete benchmarking methodology might prove very enlightening, it is probably more than you will want to do when you are trying to answer questions with your GIS. An alternative is to take a small subset of your database (the fewer grid cells, the better) and examine particularly sensitive procedures within individual submodels. This normally entails three separate processes: (1) developing a selection process for algorithms to test, (2) selecting spatially useful portions of the database against which to test them, and (3) determining how big these subsetted database portions need to be to test the selected algorithms. As we have seen, operators can range from local (by-cell) to global. Each of these presents a different scenario for selecting your prototype database portion.

For local operators, the selection of test or verification prototype databases is relatively easy. It merely requires that the cells in question represent important categories or values within your model. For focal operators, it is best that cells be selected so that some examination cells be selected from within and some from outside the near neighborhood used for the algorithmic implementation (Figure 9.1). In this way, you can determine if any cells outside the neighborhood are being inadvertently affected by the program in addition to evaluating the computations within the neighborhood itself. A similar methodology should be adopted for examining zonal operators, except that two or three locations that contain both regional and extraregional portions should be used where fragmented zones (regions) are used (Figure 9.2). Block operators can easily be examined by selecting a set of grid cells that subtends the block under consideration. I like to look at two adjacent blocks, however, just to be sure that the process is operating properly from one block to another (Figure 9.3).

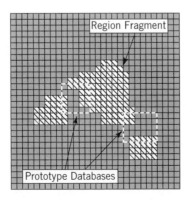

Figure 9.2 **Selecting a subset for evaluating correctness of your zonal function algorithms.**

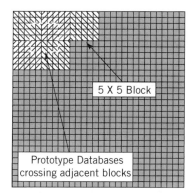

5 X 5 Block

Prototype Databases
crossing adjacent blocks

Figure 9.3   **Selecting a subset for evaluating correctness of block function algorithms.**

Among the most difficult operators for which to decide on prototype grid cells are the global operators, because they operate throughout an entire grid theme, because they may be relatively simple or highly complex, and because highly variable database grid cell values can also have an impact on the way the algorithms perform. In these circumstances, I usually try to select prototype grid cells by starting at the target cell and moving outward in all directions for at least three or four grid cells (Figure 9.4). If you encounter no problems at this level, you can normally be fairly comfortable about the performance of the code, especially where accumulated errors will likely be encountered through the progressive movement of the program.

No matter the types of operators selected for examination, the verification proceeds by first reproducing the process manually. For example, if you intend to examine a local operator—say, an overlay of two grid themes using multiplication—a sample of a contiguous coregistered 4×4 matrix of cells from each multiplied theme will suffice for selection. Comparing those selected cells from the GIS output with their analog equivalents should yield identical results (Figure 9.5). Any of the operators can be examined in this manner, although more complex algorithms may require more time than do simple local operators.

Let us say, however, that your purpose is to evaluate the precision (repeatability) of the algorithm. If you are satisfied that the algorithm is operating properly when compared with the analog manipulation, you can then evaluate the repeatability by performing it twice, finally performing a subtractive overlay operation on the results of both iterations. If they are consistent, the results should yield a set of zeros. Although it may seem that such an operation is rather wasteful, you might be surprised at how often there are grid cells that are not zeros.

Another important aspect of model verification is that of evaluating the appropriateness of how the algorithms themselves are applied to the model. As you remember, a model is an ordered sequence of map operations. The order of such operations is often crucial to the proper modeling of your variables. Reclassification operations

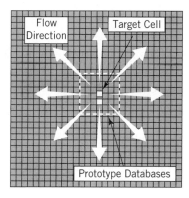

Flow Direction

Target Cell

Prototype Databases

Figure 9.4   **Selecting a subset for evaluating correctness of global function algorithms.**

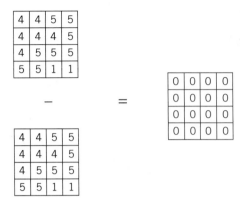

Figure 9.5 Comparing the results of analog operations with those of the computer. The results should be the same. In fact, if these were overlaid through a subtractive method, the output should be a set of zeros.

often require several steps, each of which must be performed in the correct sequence for the answer to be correct. This is particularly important if your raster GIS does not carry the legend forward as these operations are employed. This is much like the problems of distributive properties of algebraic equations, in which some sequences contained within parentheses must be performed first for the equation to achieve the correct results.

An example of how this can be problematic is the process of reclassification through the use of a renumbering command. Let us say, for example, that you are trying to reclassify some grid cells from specific land use types to a more generic one. In this case, you are trying to change *barley, wheat, oats, potatoes, soybeans,* and *beets* into a more generic set of classes called *row crops* and *grain*. Let us further say that *barley* is assigned a nominal class of *1, wheat* is *2,* oats is *3,* and *potatoes, soybeans,* and *beets* are *4, 5,* and *6,* respectively. If you issue a set of commands that change *potatoes, soybeans,* and *beets* (values *4, 5,* and *6*) to a value of *1,* you now have a grid with values of *1, 2,* and *3.* Unfortunately, the value of *1* used to be applied to *barley.* Now your grid shows *2* representing *wheat,* and *3* representing *oats,* as before. But the value of *1,* which used to represent only *barley,* now represents *wheat* and all *row crops.* If you continue your renumbering process so that a nominal value of *2* is meant to represent all grains, you will be renumbering all the remaining values (*1, 2,* and *3*) until they will all be reclassified as *grain.* What you will obtain is a map where all values will have the same number, *2* (Figure 9.6). You have effectively mixed your classes so that they are unrecognizable. With most professional GIS software, this is no longer as problematic, because you are changing the category values (i.e., *barley, oats,* and *wheat* will be now be called *grains,* and *potatoes, soybeans,* and *beets* will be reclassified as *row crops.* The problem of logical assignments can still be replicated, however, if you are not careful about how you choose your categories. To check these, you should, as before, select a small portion of your grid theme cells and perform the operation manually so you know what at least a portion of your map should look like.

A more universal problem in logic has to do with using the numerical scales improperly. For example, you may find yourself solving a problem by multiplying nominal land use values, such as the numbers *5, 10,* and *15* representing *urban, agriculture,* and *vacant land,* respectively, by ratio elevation values of 10, 20, and 30 feet. The results of such operations are often aesthetically pleasing and may even represent distributional patterns that look deceptively reasonable. What you would obtain are values, for example, of *500,* representing the multiplication of 10 feet by 5, which represents only a nominal category of *urban.* To test your logic, you must ask yourself what the value of *500* represents, especially because it can easily be misconstrued as an ordinal value. Does it mean that you have urban land use at elevation values of 10 feet? Or does it mean you should probably have applied an alternative

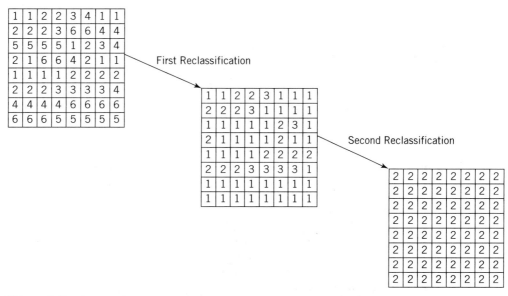

Figure 9.6   Successive renumbering operations or reclassification operations can result in unexpected results.

approach, such as applying a logic operator such that polygonal values of 10 feet are combined with polygonal values of *urban* by applying a logical "and" value to them? Thus, you are saying that you are selecting all polygons that share these two common categories rather than risking the problem of mathematical impropriety. Again, some raster GIS software has some safeguards against such illogical approaches, but you might be surprised at how many ways you can make such mistakes and how difficult they are to find after the model is complete. I recommend that you test your logic—again, by performing the operations manually on a small portion of your grid cells at each step in the modeling process. Then ask yourself the same questions we asked from our example above.

Model Validity

A more difficult and at least equally important operation in accuracy or correctness assessment is that of determining whether the model sufficiently mimics the way in which the real world it is meant to represent really operates. Bearing in mind that GIS models, like all models, are not miniaturizations of reality but rather simplifications, it is necessary to know what the underlying assumptions are before testing it. It further assumes that our understanding of the modeled environment is sufficiently complete to know how it should look and operate. If this is the case, then an analysis of the validity is a simple extension of accuracy assessment in that the logic, constraints, and mathematics of the model are assumed to be correct. So as long as the computer programs provide the answers that your logic tells you are correct, then it is a valid model.

However, this form of model validation is tantamount to circular reasoning. In essence, it says: "Assuming my logic is correct, and the algorithms correctly implement this correct logic, then the model is valid." Unfortunately, most spatial environments are far too poorly understood to make the initial assumption of correct logic. Perhaps a better way of asking the validity question is as follows: "Is the model actually modeling what we think it is?" We have already seen that even individual algorithms can be substantially wrong in how they model something as seemingly straightforward as a viewshed (Fisher 1996).

In his examination of the viewshed algorithm, Fisher (1996) employed possibly the most desirable techniques for analyzing the validity of complete models—field validation. In older remote sensing parlance, this is called ground truth, although the word *truth* does not really apply. Within GIS modeling, the purpose is not to show how satellite data act as surrogates for real data on the ground but rather how the GIS model predicts current or future conditions (in the simplest case) (Coulombe and Lowell 1995) or how effectively it prescribes the best solutions (for prescriptive models). But exactly how does one go about using field data for verifying a GIS model?

If the predictions are temporally fixed and descriptive—for example, if you are trying to show that steep, south-facing slopes in wet climates are generally less stable than gentle, north-facing slopes in temperate climates—site visitation will work. In this scenario, just visiting randomly selected points on both sites will readily illustrate the validity of this very simple model. It is also relatively easy if the model is descriptive of conditions that should be existing within relatively short time frames of any dynamic processes being used. For example, if a dam is to be built and your task is to model where the water will go when it is complete, it is a pretty simple task to identify where the water went on the dam's completion and to compare this to the prediction. This approach has been applied to predictive models of bird habitat (Nelson and Lunetta 1987). Among the most detailed and most successful published examples of field validation examines the ability of a GIS model to predict appropriate locations for the Florida scrub jay by using observational data of actual bird locations (Duncan et al. 1995).

In our LESA GIS model, our mission was to develop a model that permitted the local planners to have control over the preservation of the best agricultural lands. This was to be accomplished by restricting permits for alternative uses. One approach to model validation employed the use of surrogate field data—in this case, building permits—to represent successful application of the model. In areas where the LESA model predicted a higher likelihood of nonagricultural uses, there should be a higher percentage of permits for nonagricultural uses. This was employed in a preliminary test of the technique (Lucky and DeMers 1986–1987). To some degree, this is a measure of the acceptability of the model because it defines whether the county planners' decisions are consistent with the results, as much as whether the results themselves are consistent with the design criteria of the model. It does provide one method of analyzing results of the model itself where it might otherwise go unvalidated.

Although highly preferred as a method of validity assessment of GIS models, the field sampling methodology does have some serious limitations. First, it is often costly and time consuming to sample and analyze field data, especially if the region being modeled is large. Second, there are many scenarios, such as predictions of habitat use by wildlife, that prohibit such an approach. Not many situations exist such as those in Florida (Duncan et al. 1995), where wildlife live within predicted habitat zones, to test the hypotheses devised by the GIS model. A prediction of appropriate sites for relocating uprooted wildlife from a region, for example, assumes that there are eventually going to be actual specimens introduced to evaluate the effectiveness of the predictions. Finally, in many cases, although we might be able to sample future conditions within dynamic models such as forest fire models (Carrara et al. 1996; Liu 1998; Yuan 1994, 1997), waiting until after the actual process is complete pretty much limits the utility and acceptability of the model because the purpose is to predict forest fire damage (the descriptive part of the model) before it happens and to provide remediation (the prescriptive part of the model). In such situations, alternatives to field sampling or other direct examinations are necessary.

One effective alternative to field checking model validity is the use of validation sets or predefined results. Very few circumstances allow us to have readily available cartographic output for testing GIS models. This is primarily because if we knew the answers, we would probably not need the GIS model in the first place. Stoms (1996) suggested an alternative technique that shows promise for both remote sensing and GIS: the use of "maplets," or detailed maps of small areas, as an alternative to field checking. His approach compares detailed small-area maps for which composition, heterogeneity, and individual map unit accuracies can be compared with those of larger-area maps. In the context of GIS, one approach to employing this technique is to construct a detailed small-area prototype GIS model that allows for intensive validation prior to constructing the general model. If the larger area contains several radically different scenarios or environmental conditions, each could be prototyped in this manner. If each prototype tests out as valid (or at least reasonable), the general model can probably be considered to perform the same way. If, on the other hand, one or more subareas do not seem to perform as expected, some model modification is probably warranted.

Unfortunately, although small-area prototyping to produce validation sets is a reasonable approach, its major drawback is time constraints. An alternative to this approach is a spatial version of jacknifing, in which a portion of the general GIS model is extracted for separate modeling using the same formulation and implementation. This can then be checked against the remainder to determine whether the model performed as expected.

Such an approach can also be applied by selecting temporal rather than spatial samples from our database. Boerner et al. (1996) used this approach for predicting land use change from one period to another within a GIS. Employing a Markov Chain Model, they established a set of rules for each of three time periods, as well as for the entire beginning and ending time period. These rules were then checked for their ability to predict changes in the database.

Although this latter approach allows us to evaluate model consistency (or precision), essentially acting as a method of producing replicates, it cannot be expected to validate a model's ability to reproduce the underlying processes in the real world. To exactly measure the ability of a model to simulate real-world scenarios, we still need to know how the real world functions. In situations in which this information is lacking, the model must stand alone as a working hypothesis that will ultimately have to be tested by evaluating what actually happens in the field.

The methods described so far have focused on the comparison of map output to spatial patterns that appear either in subsets of the GIS or as spatial patterns on the landscape. Examination of the validity of a model is not restricted to this map comparison approach. Instead, the modeler can use many of the readily available statistical tests to evaluate a model. Because this is not a textbook on statistical testing, I will not endeavor to cover the entire realm of possible tests, but a couple of examples might be instructive.

Let us say, for example, that your GIS model is designed to model the dispersal of plants within a variable environment that includes a range of soils capabilities and obvious topographic features that are likely to focus or clump the seeds or other propagules. A random selection of grid cells could be applied to develop a regression model wherein the number of newly predicted plant locations can be regressed against the measurable soils and elevation values. You might even have applied such a regression model before you formulated your model to ascertain that such relationships exist. If, on applying your postmodeling regression, you find that the model distributed new plants uniformly, it should be pretty obvious from your regression that the model does not properly predict the new locations. You could also apply this latter approach by using a logistic regression applied to randomly assigned grid cells, especially where you are trying to predict point patterns.

Some forms of statistical tests are readily useful for testing within- and between-group variation from the means. Small subsamples of your model response could be subjected to analysis of variance to ascertain if there is high variability in regions that the model should have produced relatively low variability. In circumstances in which this is true, you might also apply some form of principle component analysis to your model to determine if one or more of your submodels is placing an inordinate weight on the outcome of the model. No doubt there are many more statistical tests that can be applied. I encourage you to use them whenever practical to evaluate your models, especially when the outcome involves potentially litigious situations based on decisions made from the model results.

PARSIMONY

Before we move to the topic of model acceptability, it is important to remember that there are elegant models and not-so-elegant models that can achieve the same results. The term *parsimony* is the section head here because of its direct linkage to logic. A parsimonious model is one that achieves correct results with the fewest necessary steps and the least amount of computation time. In GIS modeling, there are many ways to achieve the same results, especially for models that are meant to represent reality. As in mathematics, we find that a general rule of thumb is that we should not create a complex model when a simple one is available to us. This is also true of computer programming, in which some algorithms are more efficient and produce less code yet still achieve the same results.

To my knowledge, there has been little or no scientific literature to guide us through an examination of GIS model parsimony. For that reason, you should recognize that what is written here is largely my own private musings, based on my own teachings and research. If you find yourself asking whether your model is the most elegant, least confusing, or most straightforward approach possible, then I have achieved the desired result—getting you to think about this as you use your model. Let me begin by providing two basic lists of what I perceive to be first, the importance of parsimony in GIS modeling, and second, the things we need to examine.

Importance of Parsimony

1. The more parsimonious the model, the easier it is to explain to your client (or to yourself).

2. The more parsimonious the model, the easier it will be to check for correctness.

3. The more complex the real-world setting, the more important it is that the model demonstrate parsimony.

4. The more parsimonious the model, the easier it will be to refine and expand.

Beginning with item 1 above, it is important to remember that most clients are not familiar with the details of GIS modeling. If they were, they would probably not be using your services. The simpler the model formulation and flowchart, the less time you will have to spend explaining individual portions of the model because there will be fewer of them to explain. This is the same for item 2, in that fewer steps means fewer things that can go wrong and fewer steps have to be checked for correctness. Item 3 seems reasonable to me, although it is only a working hypothesis. A simpler model does not imply here that we have eliminated necessary steps in the process, only that redundant, meaningless, or inefficient ways of achieving the same result are eliminated. For example, why peform a series of map overlay operations when a simple reclassification will achieve the same results?

Finally, item 4 seems pretty obvious, especially if your client is a GIS neophyte. We have examined the LESA model several times in this text. The reason this planning model was developed as an additive linear model is that it provides an easily understood method of achieving rankings for potential conversion of agricultural lands to nonagricultural uses. Although a model that attempts to incorporate high-level differential calculus may more accurately represent the real world, it is not likely to be of much use if the clients do not understand what it is doing—but this is a topic we will cover a little later in this chapter. For now, we move on to ways of measuring parsimony by examining some ways of measuring it:

Ways of Measuring Parsimony

1. Number of steps
2. Simplicity of steps
3. Amount of computation time
4. Ease of comprehension
5. Number of iterations
6. Ratio of parsimony to model thoroughness

There is no more guidance from the literature as to how to *measure* GIS model parsimony than there is about parsimony at large. The above simple list should give us some guidance, however. The first of these methods is perhaps the most obvious but is not necessarily the best method of measurement. Certainly, if four people provide a GIS model, each producing the desired result, the model with the least number of computational steps would be considered to be the best as measured by parsimony. This relates to item 2, although sometimes in an inverse relationship. In some cases, simple steps can be used in place of more complex ones, such as reclassification rather than multiple overlay operations, thus actually reducing the overall number of steps necessary to produce the desired output. Alternatively, multiple simple steps may actually increase the total number of steps, and their number could easily be reduced by the use of a more complex single operation. The single complex operation may be more elegant but may also have the negative effect of requiring more time to explain. In this case, I would usually defer to item 1. There comes a time when not every step needs explaining, especially if it is a well-defined and well-established approach, such as using a neighborhood function rather than multiple reclassification steps. If there is still uncertainty as to which of these to use, item 3 might be more appropriate. If the time it takes the model to run using a complex operator is substantially longer than for several simple steps, then time is saved by multiple steps. Given the speed of most computers, this may not be a major issue unless there are many iterative runs of the model necessary to achieve the final results. In such cases, the model may take many days to run using complex algorithms that must be applied multiple times, especially if there is a need for human intervention between each step.

This in turn leads us to another method of evaluating parsimony—that of understanding the model. If there are multiple iterations with human input each time, simpler steps generally yield greater understanding for the operator, and there is less likelihood for mistakes during one or more of the iterations. This is probably more important than whether the client understands each step in the model, because the client is not as likely to want to know all the details. It has been the experience of many professional GIS applications builders that the client wants reasonable and timely results far more than a detailed explanation of what the program is actually doing. Of course, we are still assuming that the modeler is attempting to maintain some degree of professional integrity during the modeling process.

Given the possibility that our models, especially prescriptive models, are likely to involve several or even dozens of iterations to adjust to scenarios and situational changes, a reduction in the numbers of such iterations is also an important aspect of model parsimony. If, as in our LESA model, the planners simply want a general model of good places for all nonagricultural uses, as opposed to specific models for each, then the number of iterations can be reduced or even eliminated altogether with a more general descriptive model. An alternative is to create a separate model for each case. In this way, your general model is kept simple, and each new situation or scenario can be considered as a separate model. This has some real advantages in terms of the GIS use as a decision tool. Rather than having the model test all possible cases, it is simply developed anew with specific situational data added as needed. Under such circumstances, the GIS modeler and the GIS operator need to work closely, but the simplicity is maintained.

Our final factor for parsimony acknowledges the close relationship between the elegance of the model formulation and its ultimate goal of producing valid results. A simple model may very well produce a quick response, but if the results are meaningless, then validity must take precedence over elegance. Having said this, we need to recognize, again, that the desire of the clients to get a rapid response that they can easily explain to their superiors may force you to abandon this rule of thumb as well. Unfortunately, modeling is not done in a political vacuum in many real-world settings. You must use your best judgment as to your willingness to provide a less than perfectly valid output because of the time demands of the client.

Ultimately, the measurement of GIS parsimony, like the building of GIS models itself, is as much art as science. It requires a good grasp of geographic concepts, a thorough knowledge of the subject matter, insight, creativity, and a good grounding in spatial analysis. All of these, in turn, benefit from experience. The more experience the modeler has, the more adept he or she should be at producing effective GIS models with as little waste as possible. As your experience in GIS modeling grows, you should make every effort to become better at making tighter models rather than just bigger models. Stated in the reverse, my advice would be that you try to resist the temptation to impress people with your ability to create higly complex models when they are not necessary.

MODEL ACCEPTANCE

I have already alluded to some of the issues regarding model acceptance by examining issues of model correctness, validity, and parsimony. In fact, all of these issues play some role in the acceptance of the SIP in some cases, and absolutely none in other cases. This may seem an unusual statement, given the amount of material provided on these important subjects, but there are situations when the final output from GIS analysis is merely meant to provide documentation and justification for clients to make a decision that they have already determined. In such situations, the GIS modelers must again examine their own professional standards before either accepting the task or delivering the product. The next paragraphs all assume that your clients are looking for real, correct, valid, and/or understandable results.

The acceptance of a GIS modeling project is the final and possibly the most important step in the process. If you are working for clients, their payment to you may very well hinge on whether they accept what you provide them. Although many clients are content if the model produces the results for which they have contracted with you, this will become less often the case as more and more become GIS literate. Model verification is still likely more important to you as the modeler and so is not as essential to model acceptance. How the model simulates the environment that you model will be important, especially if clients are familiar with their environment.

Providing tests of the model results, especially with the clients present, is perhaps the best method of assuring them that you are providing a useful product. This in itself is not enough, however, especially if you are not going to be the primary operator of the model. You may be required to provide not only the model itself but also appropriate data sets and, more importantly, a GUI that requires little training for the non-GIS operator within your clients' office. The interface will, of necessity, be able to anticipate the kinds of questions the clients will regularly ask of the model and allow the model to perform its tasks to provide the answers. In some cases, you might also wish to include the capability of asking questions that are not regularly asked but whose answers may prove useful in selected circumstances. Alternatively, you may wish to anticipate such nonregular questions but not provide them unless they are specifically required within your contractual agreement with the clients. This would normally be done so that you can maintain a working arrangement with the clients to provide additional services. That leads us more into the applications design issues than this text was meant to address.

In any event, the GUI should be self-explanatory and easy to use. It should also be capable of selecting the appropriate data sets and then performing the modeling tasks without the user needing to know what is actually happening within the software. However, when you deliver the model and its GUI, it is probably a good idea to include at a minimum a set of schematics that explain, in general terms, what the model does when a certain button is pushed or a slider bar is moved. Flowcharts and model formulations should also be included in the model or application, probably as an appendix or separate document. This allows users to examine what the model is doing if they have basic questions about the results it provides. More detail is best provided only when requested.

Although we are assuming here that clients want the model to perform in valid ways, the results of modeling may not give them the answers they want. Remembering that the fundamental goal of most GIS modeling tasks is to provide spatial decision aids, you should be aware that even valid models may very well be unacceptable to clients, especially if their agenda is something other than purely rational decision making. If their decision making is based more on political motives than on spatial reality, you may need to modify some aspects of the model to include other factors or to weight some aspects more than others. Although this should have been well established at the time of model conceptualization or formulation, sometimes this is not the case, and modifications will have to be made.

Another aspect of GIS model acceptability has less to do with modeling than it does with timing. If you provide a model of potential landfill sites after those sites have already been purchased for alternative uses, your model may lack the timeliness necessary to be of use as a decision tool for clients. Additionally, if the model works only with data that are not as yet available, this, too, will likely negate its utility and thus its acceptability to clients. This should suggest to you that time constraints on model delivery date be clearly established well in advance of the modeling process itself.

This latter issue also suggests one final factor that may cause difficulty in getting the client to accept your final product—that of missing data. We have already discussed ways to mitigate this problem by assigning surrogates whenever possible, or placing aspatial operators in their place. There are many situations, however, when certain factors of your model are, and will remain, absent. What do you do here? There are no pat answers to this seemingly simple question. You may have noticed from the Williams (1985) article that there are a couple of missing factors for that model. Although the LESA model was developed as an academic exercise, it could just as easily have been created for a real client because it did use real data in a real study area. The first suggestion I might give is this: Do not hide the fact that data are missing. When real decisions are being made, especially if they involve real property, there is always the possibility of litigation. Many GIS models include somewhere, either within the documentation or as a separate statement, a statement that there are data missing from the model and that you as the modeler are not responsible for decisions made on the basis of the model or, more specifically, largely on the basis of factors for which there are no data.

There should be a way of avoiding this problem before it happens. As the modeling process takes place, it is unlikely that you will not become aware, well before the model is complete, that some data are going to be lacking. As soon as you are aware of the problem, you should notify the clients and ask them for suggestions before you proceed. This may damage your credibility to some minor degree, but it is much less likely to cause problems than is waiting until delivery of the final product. If clients are made aware of the missing data, they may be able to provide essential information that will assist you in creating work-arounds, or even surrogates. This should also suggest to you that the conceptualization and formulation of the model, prior to implementation, is the best time to make these discoveries. Many clients, especially GIS neophytes, will view a GIS as a panacea. You, as a modeler, should be well aware that it is not. It is best to dispel this myth early on and proceed to provide the most correct, most valid, most elegant model you can within the limits of the software, your knowledge of the environment being modeled, and the best available data. As long as you keep your goals of providing decision support (not decisions) to clients and keep them informed about the possible difficulties or limitations, you should be able to provide some of the most useful decision support available for spatial decision making.

Chapter Review

A geographic information system (GIS) model is of little utility unless it provides for decision support for the client, even if you are the client. Decisions from a GIS model require that the software code correctly implement the algorithms, that the algorithms substantially resemble the attributes and processes of the environment being emulated, and that the model be acceptable both in form and in substance to the client. This chapter considered two basic properties of model correctness—analysis of whether the model can be verified computationally and whether it is valid as a correct model of the environment. Additionally, it examined the acceptability of the model to the client both as a tool that is easy to use and one that provides answers that support the decision-making process.

The verifiability of a GIS model is a measure of how accurately the model implements the accepted algorithms for the internal computational tasks it normally performs. This process requires three steps that should be evaluated prior to testing: determining what algorithms are to be tested, selecting portions of the database that are likely to demonstrate the effects of error, and determining the correct size of the subset to be examined. A benchmarking procedure is conducted either by performing the tests manually and comparing them with their automated counterpart or, if for examining model consistency, by performing the tests several times on a preselected subportion of the database and overlaying the results using a subtractive process. In the latter case, if the results are consistent, they should yield a grid composed entirely of zeros.

Validation of a GIS model is a measure of how well the actual environment and its processes are represented by the model and its algorithms. This process may require field checking of the results of real processes, or it may employ a subsetting of the model in two different locations to test its ability to operate in different environments.

Assuming the client wishes the model to be consistent with real parameters and real-world settings rather than to be only politically expedient, its acceptability may hinge largely on its ability to successfully reproduce reality, thus providing correct answers to decision makers. This is not enough to make a model acceptable, however. The model must also be simple enough to be explained, at least in general terms, to the client. In general, the simpler the model is to explain, the more likely it is to be accepted by the client, especially if the client is not GIS proficient. The model must also include an appropriate graphical user interface if the client or the client's GIS operators are to perform the analysis with the software. Only when the final model is in a form that is acceptable to the client is the modeling process over.

Discussion Topics

1. Within the context of geographic information system (GIS) modeling, what is the difference between model verification and validation?

2. Explain some basic approaches to model verification mentioned in this text. Can you suggest other means for performing each? Begin compiling a set of references to literature that details other approaches.

3. If you do not know the algorithm for a particular operation inside your raster GIS, suggest ways that you might figure this out, particularly if the algorithm is proprietary and the vendor will not explain it to you. Hint: Think of the definition of the term *reverse engineering*.

4. Explain some of the basic approaches to model validation mentioned in your text. Can you find concrete examples in the literature other than those list in your text? Begin compiling a set of references that detailed other methods that are not included.

5. Briefly describe what is meant by *GIS model parsimony*. Why do we need to look at model parsimony anyway?

6. What are the basic measures of parsimony mentioned in this text? Discuss these with your classmates or other professionals. Are there some missing?

7. Under what circumstances might a client be willing to accept a GIS model even if it is proven to be unverifiable or invalid? Can you detail specific cases in which you have observed this?

8. If you have created a verifiably valid model for a client, why might the client still not be inclined to accept it and pay you for your services?

9. Under what circumstances might you want to omit certain functional capabilities from a GIS model?

Learning Activities

1. If you have not already done so, download the LESA (Land Evaluation and Site Assessment) database from the Wiley Web site: www.wiley.com/college/geog/ demers314234/ Select three or four simple algorithms to test and a small section of the database against which you will verify that the algorithms are correctly

implemented. In your selection, try to include at least a buffer and one overlay method to test. Now perform the manual operations to yield a truth set against which you will test the algorithm. Use your software to perform the operations on the entire database. Now test to see that the results for the selected grid cells you evaluated manually are identical to those of the output from your GIS package.

2. Using the LESA database, perform a buffer around the waterlines grid using whatever values you wish. Call the results test 1. Now perform the same operation again on the same waterlines grid. Call that result test 2. Now perform a subtractive overlay operation. What was the result? Suggest other types of overlay that could also be applied to examine the repeatability of the operation.

3. Using the LESA database and one or more of the methods of model validation discussed in this chapter, make a determination of the logical validation of the model. To do this, you may need to make some assumptions about how things work in Douglas County, Kansas. Alternatively, examine some models from the literature and suggest how you might test these using the methods suggested in this text. If you have the time, try to obtain the databases from the authors to see if you can test their validity.

4. Create a flowchart for the site assessment portion of the LESA GIS model for Douglas County, Kansas. Compare your flowchart to those of others in your class or organization. Run the models and test them for parsimony.

5. With your LESA database and software, create a graphical user interface that will allow the planners to perform an iterative situational analysis of the site assessment portion of the LESA model. Given time constraints, you may not be able to complete such a task for all operations or for all possible situations. To simplify the task, select only two or three possible land uses and focus on a single portion of the database.

REFERENCES

Agee, J.K., et al. 1989. "A Geographical Analysis of Historical Grizzly Bear Sightings in the North Cascades." *Photogrammetric Engineering and Remote Sensing* 55(11):1637–1642.

Algarni, A.M., 1996. "A System with Predictive Least-Squares Mathematical Models for Monitoring Wildlife Conservation Sites Using GIS and Remotely Sensed Data." *International Journal of Remote Sensing* 17(13):2479–2503.

Aspinall, R.J., 1994. "Exploratory Spatial Analysis in GIS: Generating Geographical Hypotheses from Spatial Data." *Innovations in GIS* 1:139–147.

Band, L.E., 1989a. "Automating Topographic and Ecounit Extraction from Mountainous Forested Watersheds." *AI Applications in Natural Resource Management* 3(4):1–11.

Band, L.E., 1989b. "Spatial Aggregation of Complex Terrain." *Geographical Analysis* 21(4):279–293.

Band, L.E., 1989c. "A Terrain-Based Watershed Information System." *Hydrological Processes* 3(2):151–162.

Band, L.E., 1993. "Extraction of Channel Networks and Topographic Parameters from Digital Elevation Data," pp. 13–42. In *Channel Network Hydrology,* Bevin, K., Ed, New York: John Wiley & Sons.

Battad, D.T., 1993. "Integration of Geographic Information Systems with Simulation Models for Watershed Erosion Prediction," Ph.D. dissertation, Texas A&M University, DAI, vol. 54–11B, p. 5468.

Batty, M., and Xie, Y., 1994. "From Cells to Cities." *Environment and Planning B: Planning & Design* 21:531–548.

Berry, J.K., 1997. *Spatial Reasoning for Effective GIS,* New York: John Wiley & Sons.

Berry, J.K., 1993. "Cartographic Modeling: The Analytical Capabilities of GIS," pp. 58–74, In *Environmental Modeling With GIS,* M.F., Goodchild, B.O. Parks, and Louis T. Steyaert, Eds., New York: Oxford University Press.

Berry, J.K., 1987. "Fundamental Operations in Computer-Assisted Map Analysis." *International Journal of Geographical Information Systems* 1(2):119–136.

Berry, J.K., 1995. *Spatial Reasoning for Effective GIS.* GIS World Books. Fort Collins, Colorado.

Boerner, R.E.J., DeMers, M.N., Simpson, J.W., Artigas, F.J., Silva, A., and Berns, L.A., 1996. "A Markov Chain Model of Land Use Inertia and Dynamism on Two Contiguous Ohio Landscapes." *Geographical Analysis* 28(1):56–66.

Brown, S., Schreier, H., Thompson, W.A., and Vertinsky, I., 1994. "Linking Multiple Accounts with GIS as Decision-Support System to Resolve Forestry Wildlife Conflicts." *Journal of Environmental Management* 42(4):349–364.

Burrough, P.A., and McDonnell, R.A., 1998. *Principles of Geographical Information Systems,* New York: Oxford University Press.

Carrara, P., Madella, P., Miuccio, A., and Rampini, A., 1996. "GRID: A Geographic Raster Image Database to Support Fire Risk Evaluation in Mediterranean Environment," pp. 289–300. In *Courses and Lectures—International Centre for Mechanical Sciences.* New York: Spinger-Verlag.

Carver, S.J., 1991. "Integrating Multi-Criteria Evaluation with Geographical Information Systems. *International Journal of Geographical Information Systems* 5(3):321–339.

Chang, K., Verbyla, D.L., and Yeo, J.J., 1995. "Spatial Analysis of Habitat Selection by Sitka Black-Tailed Deer in Southeast Alaska, USA." *Environmental Management* 19(4):579–589.

Chase, S.B., 1991. "The Integration of Hydrologic Simulation Models and Geographic Information Systems," Ph.D. dissertation, University of Rhode Island, DAI, vol. 52–08B, p. 4354.

Childress, W.M., Rykiel, Jr., E.J., Forsythe, W., Li, B., and Wu, H., 1996. "Transition Rule Complexity in Grid-Based Automata Models." *Landscape Ecology* 11(5)257–266.

Chrisman, N.R., 1997. *Exploring Geographic Information Systems,* New York: John Wiley & Sons.

Clark, K.C., 1999. *Getting Started with Geographic Information Systems.* Englewood Cliffs, NJ: Prentice-Hall.

Clark, J.D., Dunn, J.E., and Smith, K.G., 1993. "A Multivariate Model of Female Black Bear Habitat Use for a Geographic Information System." *Journal of Wildlife Management* 57(3):519–526.

Costanza, R., and Maxwell, T., 1991. "Spatial Ecosystem Modelling Using Parallel Processors." *Ecological Modelling* 58:159–183.

Coulombe, S., and Lowell, K., 1995. "Ground-Truth Verification of Relations between Forest Basal Area and Certain Ecophysiographic Factors Using a Geographic Information System." *Landscape and Urban Planning* 32(2):127–136.

Coulson, R.N., Folse, L.J., and Loh, D.K., 1987. "Artificial Intelligence and Natural Resource Management." *Science* 237:262–267.

Cromley, R.G., and Hanink, D.M., 1999. "Coupling Land Use Allocation Models with Raster GIS." *Journal of Geographical Systems* 1(2):137–153.

Davis, J.R., 1981. "Weighting and Reweighting in SIRO-PLAN." Canberra: CSIRO, Institute of Earth Resources, Division of Land Use Research, Technical Memorandum 81/2.

DeMers, M.N., 2000a. *Fundamentals of Geographic Information Systems,* 2nd ed.: New York: John Wiley & Sons.

DeMers, M.N., 2000b. *Exercises in GIS.* New York: John Wiley & Sons.

DeMers, M.N., 1992. "Resolution Tolerance in an Automated Forest Land Evaluation Model." *Computers, Environment and Urban Systems* 16:389–401.

DeMers, M.N., 1989. "Knowledge Acquisition for GIS Automation of the SCS LESA Model: An Empirical Study." *AI Applications in Natural Resources* 3(4):12–22.

DeMers, M.N. 1985. "The Formulation of a Rule-Based GIS Framework for County Land Use Planning, Lawrence, Kansas," unpublished Ph.D. dissertation.

DeMers, M.N., Simpson, J.W., Boerner, R.E.J., Silva, A., Berns, L.A., and Artigas, F.J., 1996. "Fencerows, Edges, and Implications of Changing Connectivity: A Prototype on Two Contiguous Ohio Landscapes." *Conservation Biology* 9(5):1159–1168.

Desmet, P.J.J., 1997. "Effects of Interpolation Errors on the Analysis of DEMs." *Earth Surface Processes and Landforms* 22:563–580.

Duncan, B.W., Breininger, D.R., Schmalzer, P.A., and Larson, V.L., 1995. "Validating a Florida Scrub Jay Habitat Suitability Model, Using Demography Data on Kennedy Space Center." *Photogrammetric Engineering and Remote Sensing* 61(11):1361–1370.

Dunn, W.C., 1996. "Evaluating Bighorn Habitat: A Landscape Approach." Department of Game and Fish, State of New Mexico, Technical Note 395, BLM/RS/ST-96/005+6600.

Eck, J.E., 1998. "What do Those Dots Mean? Mapping Theories with Data," pp. 379–406. In *Crime Mapping & Crime Prevention. Crime Prevention Studies* (vol. 8), D. Weisburd and T. McEwen, Eds., Monsey, NY: Criminal Justice Press.

Edwards, B. 1979. *Drawing on the Right Side of the Brain.* New York: Houghton Mifflin.

Environmental Systems Research Institute Staff, 1994. *Cell-Based Modeling with GRID,* Redlands, CA: ESRI.

Erdas Imagine Version 8.4 Tour Guides, Atlanta, Georgia.

Environmental Systems Research Institute, 2000. Using Model Builder. Redlands, CA.

Federal Geographic Data Committee, 1992. *Manual of Federal Geographic Data Products.* Washington, D.C.: Environmental Protection Agency, Office of Information Resources Management.

Fisher, P.F., 1996. "Reconsideration of the Viewshed Function in Terrain Modeling." *Geographical Systems* 3:33–58.

Fisher, P.F., 1995. "An Exploration of Probable Viewsheds in Landscape Planning." *Environment and Planning B: Planning and Design,* 22:527–546.

Fisher, P.F., 1991. "Modelling Soil Map-Unit Inclusions by Monte Carlo Simulation." *International Journal of Geographical Information Systems* 5(2):193–208.

Fisher, P., and Wood, J., 1998. "What is a Mountain? Or the Englishman Who Went up a Boolean Geographical Concept But Realized it was Fuzzy," *Geography* 83(3):247–256.

Forman, R.T.T., 1995. *Land Mosaics: The Ecology of Landscapes and Regions.* Cambridge: Cambridge University Press.

Gardner, M. 1970. "The Fantastic Combinations of John Conway's New Solitaire Game 'Life.'" *Scientific American* 223(4):120–123.

Gardner, M. 1971. "On Cellular Automata, Self-Reproduction, the Garden of Eden and the Game of 'Life.'" *Scientific American* 224(2):112–117.

Gros, S.L., Williams, T.H.L., and Thompson, G., 1988. "Environmental Impact Modelling of Oil and Gas Wells Using a GIS." *Technical Papers of the ACSM/ASPRS,* vol. 5:216–225.

Haddock, G., and Jankowski, P., 1993. "Integrating Nonpoint Source Pollution Modelling with a Geographic Information System." *Computers, Environment, and Urban Systems,* 17:437–451.

Harris, S., 1997. "Evaluating Possible Human Exposure Pathways to Populations Relative to Hazardous Materials Sites." Proceedings, Seventeenth Annual ESRI User Conference, Palm Springs, California.

Heuvelink, G., and Burrough, P. 1993. "Error Propagation in Cartographic Modelling Using Boolean Logic and Contuous Classification," *International Journal of Geographical Information Systems* 7(3):231–246.

Heuvelink, G., Burrough, P.A., and Stein, A., 1989. "Propagation of Errors in Spatial Modelling with GIS." *International Journal of Geographical Information Systems* 3(4):303–322.

Heywood, I., Cornelius, S., and Carver, S., 1998. *An Introduction to Geographical Information Systems,* Essex: Addison Wesley Longman.

Hilborn, R., 1979. "Some Long Term Dynamics of Predatory–Prey Models with Diffusion." *Ecological Modelling* 6(1):23–30.

Hodgson, M.E., and Gaile, G.L., 1999. "A Cartographic Modeling Approach for Surface Orientation-Related Applications. *Photogrammetric Engineering and Remote Sensing* 65(1):85–95.

Hogeweg, P. 1988. "Cellular Automata as a Paradigm for Ecological Modelling." *Applications of Mathematics and Computing* 27:81–100.

Hopkins, L.D., 1977. "Methods for Generating Land Suitability Maps: A Comparative Evaluation." *American Institute of Planners Journal* 43:386–400.

Ive, J.R., and Cocks, K.D., 1983. "SIRO-PLAN and LUPLAN: An Australian Approach to Land Use Planning. 2. The LUPLAN Land-Use Planning Package." *Environment and Planning B: Planning and Design* 10(3):347–355.

Ive, J.R., and Cocks, K.D., 1989. "Incorporating Multi-Party Preferences into Land Use Planning." *Environment and Planning B: Planning and Design* 16:99–109.

Iverson, D.C. & R.M. Alston, 1986. "The Gensis of FORPLAN: A Historical and Analytical Review of Forest Service Planning Models," Intermountain Research Station. USDA Forest Service. General Technical Report. INT-214.

Jenny, H., 1941. *Factors of Soil Formation.* New York: McGraw-Hill.

Jensen, J.R., 2000. *Remote Sensing: An Environmental Perspective.* Englewood Cliffs, NJ: Prentice-Hall.

Jenson, S.K., and Domingue, J.O., 1988. "Extracting Topographic Structure from Digital Elevation Data for Geographic Information System Analysis," Photogrammetric Engineering and Remote Sensing 54(11):1593–1600.

Johnston, K.M., 1992. "Using Statistical Regression Analysis to Build Three Prototype GIS Wildlife Models." Proceedings, GIS/LIS' 92, San Jose, ASCM-ASPRS-URISA-AM/FM, 1:374–386.

Kelly, G.A. (1955). *The psychology of personal constructs.* New York: Norton.

Kemp, K.K., 1993. "Spatial Databases: Sources and Issues," pp. 361–371. In *Environmental Modeling with GIS,* M.F. Goodchild, B.O. Parks, and Louis T. Steyaert, Eds., New York: Oxford University Press.

King, A.W., Johnson, A.R., and O'Neill, R.V., 1991. "Transmutation and functional representation of heterogeneous landscapes." *Landscape Ecology* 5(4):239–253.

Konikow, L.F., and Bredehoeft, J.D., 1978. "Computer Model of Two-Dimensional Solute Transport and Dispersion in Ground Water, USGS Techniques of Water Resources Investigations," book 7, chapter C2, Washington, D.C.: U.S. Geological Survey.

Lesser, T., Wei-Ning, X., Furuseth, O., McGee, J., and Lu, J., 1991. "Conflict Prevention in Land Use Planning Using a GIS-Based Support System." *GIS/LIS Proceedings* 1:478–483.

Leung, Y. 1988. *Spatial Analysis and Planning Under Imprecision.* Amsterdam. Elsevier Science Publishers B.V.

Leung, Y. and Leung, K., 1993. "An Intelligent Expert Systems Shell for Knowledge-Based Geographical Information Systems: I. The Tools." *International Journal of Geographical Information Systems* 7(3):189–199.

Lillesand, T.M., and Kiefer, R.W., 2000. *Remote Sensing and Image Interpretation,* 4th ed., New York: John Wiley & Sons.

Liu, P., 1998. "A Probabilistic GRID Automation of Wildfire Growth Simulation," Ph.D. dissertation, University of California, Riverside. DAI, vol. 59-09A, p. 3591 (206 pp.).

Lowell, K. 1991. "Utilizing Discriminant Function Analysis with a Geographical Information System to Model Ecological Succession Spatially." *International Journal of Geographical Information Systems* 5(2):175–191.

Luckey, D., and DeMers M.N., 1986–1987. "Comparative Analysis of Land Evaluation Systems for Douglas County." *Journal of Environmental Systems* 16(4):259–278.

Mackay D.S., Robinson, V.B., and Band, L.E., 1992. "Classification of Higher Order Topographic Objects on Digital Terrain Data." *Computers, Environment & Urban Systems* 16(6):473–496.

Mandelbrot, B.B., 1988. *Fractal Geometry of Nature.* W.H. Freeman.

Marble, D.F. 1994. "An Introduction to the Structured Design of Geographic Information Systems," pp. In *The AGI Source Book for GIS,* D. Green and D. Rix, Eds. London: Association for Geographical Information and John Wiley & Sons.

Marble, D.F., 1995. An Introduction to the Structured Design of Geographic Information Systems, pp. 31–38 in (Source Book, Association for Geographic Information). London: John Wiley & Sons, Inc.

Mark, D.M., 1988. "Network Models in Geomorphology," pp. In *Modelling in Geomorphological Systems,* New York: John Wiley & Sons.

Martin, D., 1996. "An Assessment of Surface and Zonal Models of Population." *International Journal of Geographical Information Systems* 10(8):973–989.

McGarigal, K. and Marks, B.J., 1994. FRAGSTATS, Spatial Analysis Program for Quantifying Landscape Structure, version 2. Corvallis: Oregon State University, Forest Science Department.

Mattikalli, N.M., 1995. "Integration of Remotely-Sensed Raster Data with a Vector-based Geographical Information System for Land-Use Change Detection," *International Journal of Remote Sensing,* 16(15):2813–2828.

Meaille, R., and Wald, L., 1990. "Using Geographical Information Systems and Satellite Imagery within a Numerical Simulation of Regional Urban Growth." *International Journal of Geographical Informations Systems* 4(4):445–456.

Miller, R.I., Stuart, S.N., and Howell, K.M., 1989. "A Methodology for Analyzing Rare Species Distribution Patterns Utilizing GIS Technology: The Rare Birds of Tanzania." *Landscape Ecology* 2(3):173–189.

Miyamoto, H., and Sasaki, S., 1997. "Simulating Lava Flows by an Improved Cellular Automata Method." *Computers & Geosciences* 23(3)283–292.

Muehrcke, P., and Muehrcke, J., 1998. *Map Use: Reading, Analysis and Interpretation,* 4th ed., Madison, WI: JP Publications.

Nelson, M.D., and Lunetta, R.S., 1987. "A Test of 3 Models of Kirtland's Warbler Habitat Suitability." *Wildlife Society Bulletin* 24:89–97.

Park, S., 1996. "Integration of Cellular Automata and Geographic Information Systems for Modeling Spatial Dynamics," Ph.D. dissertation, University of South Carolina, DAI, vol. 57-03B, p. 1684 (205 pp).

Parrat, L.G., 1961. *Probability and Experimental Errors,* New York: John Wiley & Sons.

Pereira, J.M., and Itami, R.M., 1991. "GIS-Based Habitat Modeling Using Logistic Multiple Regression: A Study of the Mt. Graham Red Squirrel." *Photogrammetric Engineering and Remote Sensing* 57(11):1475–1486.

Pereira, J.M.C., and Duckstein, L., 1993. "A Multiple Criteria Decision-Making Approach to GIS-Based Land Suitability Evaluation." *International Journal of Geographical Information Systems* 7(5):407–424.

Philip, G.M., and Watson, D.F., 1982. "A Precise Method for Determining Contoured Surfaces." *Australian Petroleum Exploration Association Journal* 22:205–212.

Portugali, et al. 1994, "Sociospatial Residential Dynamics: Stability and Instability Within a Self-Organizing City," *Geographical Analysis,* 26(4):321–340.

Raju, K.A., Slikdar, P.K., and Dhingra, S.L., 1998. "Micro-simulation of Residential Location Choice and Its Variation." *Computers, Environment, and Urban Systems* 22(3):203–218.

Robinson, V.B., 1990. "Interactive Machine Acquisition of a Fuzzy Spatial Relation." *Computers and Geosciences* 16(6):857–872.

Robinson, A.H., Morrison, J.L., Muehrcke, P.C., Kimerling, A.J., and Guptill, S.C., 1995. *Elements of Cartography,* 6th ed., New York: John Wiley & Sons.

Sauer, C.O., 1925. "Morphology of Landscapes," pp. 315–350. In: *Land & Life,* J. Leighly, Ed., Berkeley: University of California Press, 1963.

Schuster, S.A., 1973. "Locating Optimal Sites in Geographic Information Systems." Ph.D. dissertation, University of Illinois at Urbana-Champaign, DAI, 34-09B: 4328.

Scott, M.S., 1997. "Extending Map Algebra Concepts for Volumetric Geographic Analysis," Proceedings, *GIS/LIS '97 International Conference,* Cincinnati, pp. 309–315.

Shaffer, C.A., Samet, H., and Nelson, R.C., 1990. "QUILT: A Geographic Information System Based on Quadtrees." *International Journal of Geographical Information Systems* 4(2):103–131.

Shannon, C.E., and Weaver, W., 1949. *A Mathematical Theory of Communication.* Urbana, Il: University of Illinois Press.

Shreve, R.L., 1966. "Statistical Law of Stream Number." *Journal of Geology* 74:17–37.

Stoms, D.M., 1996. "Validating Large-Area Land Cover Databases with Maplets." *Geocarto International* 11(2):87–95.

Strahler, A.N., 1957. "Quantitative Analysis of Watershed Geomorphology." *Transactions of the American Geophysical Union* 8(6):913–920.

Takeyama, M., and Couclelis, H., 1997. "Map Dynamics: Integrating Cellular Automata and GIS Through Geo-Algebra." *International Journal of Geographical Information Science* 11(1):73–91.

Tarboton, D.G., Bras, R.L., Rodriguez-Iturbe, I., 1991. "On the Extraction of Channel Networks from Digital Elevation Data." *Hydrological Processes* 5:81–100.

Tauxe, J.D., 1994. "Porous Medium Advection-Dispersion Modeling in a Geographic Information System." Ph.D. dissertation in civil engineering. Austin: University of Texas.

Taylor, J.R., 1982. *An Introduction to Error Analysis,* Oxford: Oxford University Press.

Theobald, D., and Gross, M.D., 1994. "EML: A Modeling Environment for Exploring Landscape Dynamics." *Computers, Environment & Urban Systems* 18(3):193–204.

Thomas, E.N., 1964. "Maps of Residuals from Regression," pp. 326–352. In *Spatial Analysis: A Reader in Statistical Geography,* B.J.L. Berry and D.F. Marble, Eds., Englewood Cliffs, NJ: Prentice-Hall.

Tomlin, C.D., 1991. "Cartographic Modeling," pp. 361–374. In *Geographical Information Systems: Principles and Applications,* M. Goodchild, D. Maguire, and D. Rhind, Eds., Harlow, Essex, UK: Longman Group Ltd.

Tomlin, C.D., 1990. *Geographic Information Systems and Cartographic Modeling,* Englewood Cliffs, NJ: Prentice Hall.

Tomlin, C.D., 1983. "An Introduction to the Map Analysis Package." *Proceedings, National Conference on Resource Management Applications: Energy and Environment,* August 22–26. San Francisco: pp. 1–14.

Tomlin, C.D., and Berry, J.K., 1979. "A Mathematical Structure for Cartographic Modeling in Environmental Analysis," Proceedings, ACSM, Washington, D.C., March 18–24, pp. 269–284.

Tomlin, C.D., and Johnston, K.M., 1991. "The ORPHEUS Land Use Allocation Model." *Journal of Cross-Disciplinary Exchange of Knowledge in the Geosciences* 3(3)10–13.

Tomlin, S.M., 1981. "Timber Harvest Scheduling and Spatial Allocation," master's thesis, Forest Science, Harvard University, Cambridge, MA.

van Deursen, W.P.A., 1995. "Geographical Information Systems and Dynamic Models: Development and Application of a Prototype Spatial Modelling Language." Doctoral dissertation, University of Utrecht, *NGS 190.*

Wang, L., 1999. "Projection Systems in Geographic Information Systems (GIS): Comparing Distortion Difference Between Map Projections," Abstracts, Association of American Geographers, Honolulu.

Watson, D.F., and Philip, G.P., 1985. "A Refinement of Inverse Distance Weighted Interpolation." *Geo-Processing* 2:315–327.

Wesseling, C.G., Karssenberg, D., Van Deursen, W.P.A., and Burrough, P.A., 1996. "Integrating Dynamic Environmental Models in GIS: The Development of a Dynamic Modelling Language." *Transactions in GIS* 1:40–48.

Williams, T.H.L., 1985. "Implementing LESA on a Geographic Information System—A Case Study." *Photogrammetric Engineering and Remote Sensing* 51(12):1923–1932.

Wu, F., 1996. "A Linguistic Cellular Automata Simulation Approach for Sustainable Land Development in a Fast Growing Region." *Computers, Environments and Urban Systems* 20(6):367–387.

Yuan, M., 1994. Representation of Wildfire in Geographic Information Systems," Ph.D. Dissertation, State University of New York at Buffalo.

Yuan, M., 1994. Representation of Wildfire in Geographic Information Systems," Ph.D. Dissertation, State University of New York at Buffalo.

Yuan, M., 1997. "Use of Knowledge Acquisition to Build Wildfire Representation in Geographical Information Systems." *International Journal of Geographical Information Science* 11(8):723–745.

Zeff, I.S., 1991. A Cartographic Model for Land Use Planning in the U.S. Forest Service," Unpublished Master's Thesis, School of Natural Resources, The Ohio State University.

INDEX